Mathematical Fallacies,
Flaws, and Flimflam

© 2000 by
The Mathematical Association of America (Incorporated)
Library of Congress Catalog Card Number 99-67971

ISBN 0-88385-529-1

Printed in the United States of America

Current Printing (last digit):
10   9   8   7   6   5   4   3   2   1

Mathematical Fallacies, Flaws, and Flimflam

Edward J. Barbeau

University of Toronto

Published by
THE MATHEMATICAL ASSOCIATION OF AMERICA

SPECTRUM SERIES

The Spectrum Series of the Mathematical Association of America was so named to reflect its purpose: to publish a broad range of books including biographies, accessible expositions of old or new mathematical ideas, reprints and revisions of excellent out-of-print books, popular works, and other monographs of high interest that will appeal to a broad range of readers, including students and teachers of mathematics, mathematical amateurs, and researchers.

Polyominoes, by George Martin

The Search for E. T. Bell, also known as John Taine, by Constance Reid

Shaping Space, edited by Marjorie Senechal and George Fleck

Student Research Projects in Calculus, by Marcus Cohen, Arthur Knoebel,
    Edward D. Gaughan, Douglas S. Kurtz, and David Pengelley

The Trisectors, by Underwood Dudley

Twenty Years Before the Blackboard, by Michael Stueben with Diane Sandford

The Words of Mathematics, by Steven Schwartzman

MAA Service Center
P. O. Box 91112
Washington, DC 20090-1112
800-331-1622   FAX 301-206-9789

To
Victoria Isabelle Barbeau

FOREWORD

Mathematics is a dangerous enterprise. Through hard experience, mathematicians have learned to subject even the most "evident" assertions to rigorous scrutiny, as intuition and facile reasoning can often lead them astray. However, the impossiblity and impracticality of completely watertight arguments make it possible for errors to slip by the most watchful eye. They are often subtle and difficult of detection. When found, they can teach us a lot and can present a real challenge to straighten out.

For the mathematics teacher, one source of such errors is the work of students. While students are responsible for a certain amount of plain nonsense, a few more thoughtful ones make "good" mistakes that on the face seem to compel assent; sometimes it may only be a vague hunch of something not quite right that induces a marker to look more closely. Occasionally, a text book will weigh in with a specious result or solution. Presenting students with faulty arguments to troubleshoot can be an effective way of helping them critically understand material, and it is for this reason that I began to compile fallacies and publish them in the *Notes* of the Canadian Mathematical Society.

Almost a dozen years ago, I persuaded the editors of the *College Mathematical Journal* to publish a regular department devoted to fallacies and so provide a regular outlet for them that did not seem to exist elsewhere. The catchy title, *Fallacies, Flaws and Flimflam* came from the editors, Professors William and Ann Watkins. I hoped to challenge and amuse readers, as well as to provide them with material suitable for teaching and student assignments. This book collects the items from the first eleven years. Many of them were provided and commented upon by readers of the journal. Some items are student "howlers" gleaned from homework and examination scripts. Often these reveal the correct answer by a process that defies analysis, but sometimes the steps have a range of validity which can be an interesting exercise to determine.

Nonprofessional sources, such as newspapers, are responsible for a goodly number of mishaps, particularly in arithmetic (especially percentages) and probability; their use in classrooms may help students become critical readers and listeners of the media. Quite a few items come from professional mathematicians. While some result from oversights, others are deliberately crafted to either mystify or drive home an important point.

The reader will find in this book some items that are not erroneous but *seem* to be. These need a fuller analysis to clarify the situation. All the items, esteemed reader, are presented for your entertainment and use.

Since the fallacies and their analysis are printed together, there is a coding system to warn the reader against reading on too quickly. A heart or a spade indicates the end of a formal solution or proof. The black suits (spades and clubs, ♠ and ♣) warn the reader that some analysis is about to follow. The other suits (hearts and diamonds, ♡ and ♢) indicate the end of a line of thought, but the reader can continue to move ahead. The items are generally provided with an analysis, except where they are either completely nonsensical, the issue is clear or there is reference to the literature. I have tried to ensure that the errors in this book are all intentional, but expect that this hope is unrealistic in view of the treachery of the enterprise. Amendments and comments from the readers are most welcome.

I am indebted to David Andrews of the University of Toronto for reviewing the probability chapter, and to the editors for some suggestions. However, any mistakes — intentional or otherwise — are my own responsibility. Finally, I would like to thank my wife, Eileen, for her patience during the preparation of this book.

Ed Barbeau, September 14, 1999

Contents

Chapter 1

NUMBERS

Arithmetic is one of the first topics taught to elementary pupils. This is appropriate because of the prevalence of numbers in modern society and the need to have a numerate population. In particular, we often need to negotiate percentages and it is here that public understanding frequently falters. We begin with a few examples

1. How to get drunk and rich at the same time

The column *Money angles: where else to invest?* by Andrew Tobias in the May 17, 1993, issue of *Time* offers this advice for improving your financial worth:

> Buy staples in bulk when they're on sale . . .. Consider a family that buys one bottle of wine each week. With the 10% discount many stores offer on wine by the case, they would be saving 10% every twelve weeks—more than 40% a year, tax free and largely risk free.

One can in fact do better; increase consumption to one case per month and save 120% over a year, thus qualifying for a 20% payback from the merchant.

Submitted by Larry Zeitel of Loras College in Dubuque, IA.

2. Fifty per cent more for fifty per cent less

The Summer, 1997 issue of *Adobe Magazine* (Volume 8, Number 5), includes an advertisement for *Apple Macintosh*. In describing its Power Mac, the company makes this statement.

> In fact, Adobe® Photoshop® runs 50% faster on a Power Mac**.
> Which translates into 50% less time staring at your screen and

waiting for your computer to finish retouching photos, manipulating images or applying filters.

It would be interesting to see what would happen if it could run 150% faster. Presumably, "50% faster" means that on a Power Mac, Photoshop processes, say 1500 objects in the time that it would take to process 1000 such objects on a PC. This should correspond to $33\frac{1}{3}\%$ less time spent staring and waiting, not 50%. An alternative interpretation that would legitimate the translation into 50% less time spent would have a task taking 60 seconds on a PC requiring only 30 seconds on a Power Mac. But this is more correctly characterized as twice as fast.

Submitted by Norton Starr of Amherst College in Massachusetts.

3. Whose real world?

Inspired by the article of Underwood Dudley, *Is Mathematics Necessary?* (CMJ 28 (1997), 360–364), Elizabeth Berman Appelbaum of Shawnee Mission, KS, has a few comments on some more textbook examples of "real-world" problems.

Problem. Samantha has a total of 815 points so far in her algebra class. At the end of the course she must have 82% of the 1100 points possible in order to get a B. What is the lowest score she can earn on the 100-point final to get a B in the class?

Comment by EBA: When I taught college courses, students frequently asked me to do this kind of calculation. For example, if you get a D on each of three tests, what must you make on the final to get a B for the course? Finally, I realized the question is frivolous and stopped doing these calculations. If you get a D on three tests, you will get a D or F on the final. Scores on a test are a random variable that regresses to the mean. Your mean at a certain time is a good prediction for a subsequent test. I believe that students usually do a little worse on the final than on previous tests. In this example, Samantha's current average is 81.5%, and she should not bet on getting a B. If she really wanted a B, she should have worked harder earlier on.

Problem. Jack Glover wishes to add enough 50% antifreeze solution to 16 gallons of a 5% antifreeze solution to obtain a 20% antifreeze solution. How much of the 50% solution should he add?

Comment by EBA: I just talked to an employee at a service station. He did this kind of problem in mathematics classes, but on the job nobody does it this way. They test the antifreeze in the car for the temperature at which it provides protection. Then they add enough antifreeze to get to the desired temperature.

Problem.   A liter of saline solution contains 1% salt. How many cubic centimeters of salt must you add so the concentration is 2% salt?

Comment by EBA: This problem should be presented with weight, not volume, of salt. You can keep on adding salt, but you will not increase the volume, because the salt dissolves.

While many people use some mathematics in their jobs, Dr. Appelbaum feels that mathematics should nonetheless be taught primarily for its own sake. Applications are fine if they are simple and appealing; otherwise they should be left to an applied course. She sees the current emphasis on applications as a response to anti-intellectualism among students. When students ask what good the mathematics is, she suspects that the students are really saying that they cannot understand the subject and so hope that it is no good. She has often met people who are glad that they studied mathematics, or wish that they had studied more, but never anyone who said that they were sorry to have learned mathematics.

4. United in purpose

In an attempt to counteract a steady drop in faculty salaries, faculty members in the Los Angeles Community College District were asked to sign and forward to a member of the Board of Trustees this card:

Dear Trustee Conner:

We're united in purpose. **WE DEMAND A SALARY INCREASE.**

	Salary increase	Inflation CPI	Gain
1990	0	6	−6
1991	0	3.5	−3.5
1992	0	3.2	−3.2
1993	0	2.5	−2.5
1994	3	2.5	+0.5
1995	2.7	2.3	+0.4
			−14.3

We're tired of working in negative numbers. ◇

Following the reasoning behind the last two columns, one would conclude that in a year of 100% inflation we would experience a 100% loss of income. With the reasoning behind the total for the last column, if we had a 50% cut in pay one year and a 100% increase in the next, then we would enjoy a net salary increase of 50% over the two years.

Submitted by Bruce Yoshiwara of Pierce College in Woodland Hills, CA.

5. A case of black and white—but not so much black

Peter Rosenthal of the University of Toronto in Ontario is both a mathematician and a lawyer. Recently, he represented an applicant who made representation to Ontario Court of Justice concerning a constitutional challenge to a jury panel. In Ontario, only citizens can be selected for jury duty; it is not enough to be a permanent resident. Since the proportion of citizens that are black is less than the proportion of the population that is black, the citizenship requirement has a negative impact on the probability of choosing black persons as jurors. Rosenthal argued that the citizenship requirement was discriminatory within the meaning of the Canadian Charter of Rights and Freedoms. This application was opposed by Her Majesty the Queen, represented by a counsel for the Attorney-General of Canada, in the following lines:

> Furthermore, it is submitted that the citizenship requirement does not result in a pool of jurors in which blacks are differentially excluded to the extent that a representative jury cannot be obtained in Metropolitan Toronto. It is submitted that the difference in proportions as between black citizens and all blacks, and non-black citizens and all non-blacks (65.9% for blacks and 85.6% for non-blacks), is not such that the Applicants will be unable to have a realistic opportunity to have a panel which will include blacks. This becomes clear when one compares the percentage that blacks are of the total Metropolitan Toronto populations, 4.1%, with the percentage that black citizens are of the Metropolitan Toronto population who are citizens, 2.7/84.8 or 3.2%. In short, the citizenship requirement for jury service results in a pool for the array in which 3.2% of the available jurors are black, which is nearly the same proportion that blacks, citizens and non-citizens combined, are in the total population of Metropolitan Toronto, that is 4.1%. In fact, having regard to the difference in size of the black and non-black *non-citizen* groups, 1.4% and 13.8% respectively of the Metropolitan Toronto population, the inclusion of non-citizens in the array would probably result in *fewer* blacks being selected because a greater number of non-blacks would be available in the expected jury pool than blacks. ♣

The application failed; the decision is being appealed to a higher court.

Counsel actually negotiated the figures quite well. The proportions of residents of different categories in Metropolitan Toronto is given by the table

	Blacks	Nonblacks	
Citizens	2.7%	82.1%	84.8%
Noncitizens	1.4%	13.8%	15.2%
	4.1%	95.9%	100%

Indeed, 65.9% of 4.1 is 2.7, 85.6% of 95.9 is 82.1 and 2.7 is 3.2% of 84.8. And certainly adding noncitizens to the jury pool would bring in more nonblacks than blacks. But blacks would represent a larger *proportion,* and it is this that governs the probability.

6. Effects of changing temperature

Changes from one type of unit of measurement to another, say metres to feet, generally involve just multiplication by a constant of proportionality. Temperature is an exception, and this fact is responsible for slips in the following passages on pages 301 and 302 of the book *The time before history* by Colin Tudge ("Touchstone" edition, Simon & Schuster, Old Tappan, NJ, 1997).

Twenty years could see a 7° C (44.6° F) rise in temperature; the difference between a frozen landscape and a temperate one. [*Editor's comment:* Especially in the U.S., but not so much in Canada.]

Besides, at the trough of the last ice age, 18,000 years ago, the surface of the sea in the eastern Mediterranean is known to have cooled by more than 6° C (43° F), which in ecological terms is huge, yet the elephants and their miniaturized neighbours came through those harsh times.

Submitted by Dave Trautman from the Citadel in Charleston, SC.

7. To those that have shall be given

In 1992, it was found that employees of Clemson University who used their cars for official business were reimbursed 21 cents per mile, except for seven vice-presidents, who got 42 cents per mile. The president explained that he had decided to pay his vice-presidents a higher mileage rate because he felt that they had to use their cars more than other employees. In the ensuing uproar, this extra payment was rescinded, although only some of the vice-presidents actually had taken advantage of it.

Submitted by John W. Kenelly of Clemson University in South Carolina.

It sometimes happens that performing invalid arithmetic operations will yield a correct result. It is often a useful exercise to analyze the exact conditions under which this occurs.

8. Distributing addition over multiplication

Normally, distributing addition over multiplication will lead to error, but here is an exception:

$$(0.5) + (0.2)(0.3) = (0.5 + 0.2)(0.5 + 0.3) = (0.7)(0.8) = 0.56.$$

Indeed,

$$a + bc = (a + b)(a + c) \iff a + b + c = 1 \quad \text{or} \quad a = 0.$$

A "Quickie" by A. Wayne from Mathematics Magazine 48 (1975) 117, 122 found by K.R.S. Sastry of Bangalore, India.

9. Distributing exponents over sums

Observe that

$$\left(3 + \frac{3}{8}\right)^{2/3} = \frac{9}{4} = \frac{2}{3}\left(3 + \frac{3}{8}\right).$$

What other instances are there for which

$$\left(a + \frac{b}{c}\right)^{m/n} = \left(a + \frac{b}{c}\right)\left(\frac{m}{n}\right)$$

is true? ♣

We can construct examples by selecting rationals u and v for which $u^v = vu$. Either $v = 1$ or $v^{1/(v-1)} = u$, a rational. Let $v - 1 = 1/n$ where n is an integer distinct from 0 and -1. Then $v = 1 + (1/n)$ and $u = (1 + \frac{1}{n})^n$. For example, $n = 1, 2, 3, -2, -3$ give respectively

$$(u, v) = (2, 2), \left(\frac{9}{4}, \frac{3}{2}\right), \left(\frac{64}{27}, \frac{4}{3}\right), \left(4, \frac{1}{2}\right), \left(\frac{27}{8}, \frac{2}{3}\right).$$

Problem E1190 from the American Mathematical Monthly 62 (1955) 655; 63 (1956) 345–346.

10. An exponential mess

Using the well-known rule for multiplying numbers raised to powers:

$$n^{\frac{a}{b}} \cdot m^{\frac{c}{d}} = (nm)^{\frac{a}{b}+\frac{c}{d}} = (nm)^{\frac{a+c}{b+d}}$$

a student recently evaluated $3^{2/3} \cdot 9^{7/6}$ as $(3 \cdot 9)^{9/9} = 27^1 = 27$. You don't like this? Then do it your way and see if you can get something better.

This howler is from Eric Chandler of Randolph-Macon Woman's College in Lynchburg, VA.

11. A product of logarithms

Problem. Without tables, determine $(\log_3 169)(\log_{13} 243)$.

Solution. Let $a = \log_3 169$ and $b = \log_{13} 243$. Then $3^a = 169$ and $13^b = 243$. Therefore, $0 = 169 - 3^a + 243 - 13^b = 13^2 - 13^b + 3^5 - 3^a$. Hence $b = 2$, $a = 5$ and the answer is $ab = 10$. ♠

While the values of a and b are clearly wrong, the answer is in fact correct. To analyze the situation, note that the decreasing concave curve with equation $3^x + 13^y = 412$ intersects the rectangular hyperbola with equation $xy = 10$ in two points $(5, 2)$ and $(a, b) = (\log_3 169, \log_{13} 243) \doteq (4.7, 2.1)$. Thus among the four statements
A. $3^x = 169$; $13^y = 243$
B. $3^x + 13^y = 412$
C. $(x, y) = (5, 2)$
D. $xy = 10$
we have the logical relationships $A \Rightarrow B$, $A \Rightarrow D$, $C \Rightarrow B$, $C \Rightarrow D$, with none of the converses holding. In particular, the passage from A to B entails a loss of information, so that B can hold under conditions other than A. The result that A implies D required in the problem needs an independent argument. From the fact that $\log_m u = (\log_n u)/(\log_n m)$, it is straightforward to establish the striking result

$$\log_m u \log_n v = \log_n u \log_m v$$

for positive real m, n, u, v with m, n exceeding 1. Our particular result follows.

A generalization is given in the paper
Chris Freiling, The change of base formula for logarithms. *College Math. Journal* 17 (1986) 413.

12. A divisibility property

Proposition. *For any positive integer n, $n^n - n^2 + n - 1$ is divisible by $(n-1)^3$.*

Proof. $n^n - n^2 + n - 1 = n^2(n^{n-2} - 1) + (n - 1) = (n - 1)P(n)$ where $P(n) = n^2(n^{n-3} + n^{n-2} + \cdots + 1) + 1$. Since $n^2 \equiv n \equiv 1 \bmod (n - 1)$, we find that

$$P(n) \equiv n(n - 2) + 1 = n^2 - 2n + 1 = (n - 1)^2.$$

Thus, $P(n)$ is divisible by $(n-1)^2$ and the result follows. ♠

The result fails for $n = 3$. While it is true that $P(n) \equiv (n - 1)^2$ (mod $n - 1$), this simply says that $P(n) = k(n - 1) + (n - 1)^2$ for some integer k. This is not divisible by $(n - 1)^2$ unless k itself is so divisible. Thus $P(3) = 10 = 3 \times 2 + 2^2 \equiv 2 \pmod{2^2}$. Note however that $P(4) = 81 = 24 \times 3 + 9 = (8 + 1) \times 3^2$.

13. All perfect numbers are even

We recall that two functions f and g from **N** to **R** are related by

$$f(n) = \sum_{d|n} g(d)$$

if and only if

$$g(n) = \sum_{d|n} \mu(d) f\left(\frac{n}{d}\right)$$

where each summation is over the positive divisors of n and μ is the Möbius function ($\mu(1) = 1$; $\mu(n) = (-1)^k$ when n is the product of k distinct primes; $\mu(n) = 0$ otherwise).

Proposition. *Every perfect number is even.*

Proof. The positive integer n is perfect (i.e., equal to the sum of its smaller positive divisors) if and only if $2n = \sum_{d|n} d$. Inverting this, as above, yields

$$n = \sum_{d|n} \mu(d) 2\left(\frac{n}{d}\right) = 2 \sum_{d|n} \mu(d)\left(\frac{n}{d}\right).$$

Each summand is an integer, and we conclude that n is an even integer. ♠

Since $2n = \sum_{d|n} d$ is not a valid equation for every value of n, the Möbius Inversion Formula is not applicable. For a reference to the formula, consult, for example, G.H. Hardy and E.M. Wright, *An Introduction to the Theory of Numbers* (Oxford, 1938, 1945, 1954, 1960). It appears in Section 16.4 of the fourth edition. Another reference is W.J. LeVeque, *Topics in Number Theory*, Volume I (Addison-Wesley, 1956), page 87.

Argument constructed by Virek Mohta and communicated by Ari Turner of Los Alamos, NM

14. Why Wiles' proof of the Fermat Conjecture is false

In a *Parade Magazine* column of November 21, 1993, Marilyn Vos Savant criticizes the proof offered by Andrew Wiles of the Fermat Conjecture that for positive integers n exceeding 2, the equation $x^n + y^n = z^n$ has no solutions in positive integers. After discussing the advent of non-Euclidean geometries in the nineteenth century, she comments:

> Wiles' proof is also non-Euclidean. The chain of proof is based on hyperbolic geometry, which one of its founders himself named "imaginary geometry". Here is the crux of the matter.

She describes how the three ancient problems (duplication of the cube, trisecting an angle with compasses and straightedge, squaring a circle) were shown to be impossible of solution.

> Bearing all this in mind, what would we think if it were discovered that János Bolyai, one of the three founders of hyperbolic geometry, managed to " square the circle" — but only by using his own hyperbolic geometry? Well, that's exactly what happened. And Bolyai himself said that his hyperbolic proof would not work in Euclidean geometry.
>
> So one of the founders of hyperbolic geometry (the geometry used in the current proof of Fermat's last theorem) managed to square the circle?! Then why is it known as such a famous impossibility? Has the circle been squared or has it not?
>
> Has Fermat's last theorem been proved, or has it not? I would say it has not: if we reject a hyperbolic method of squaring the circle, we should also reject a hyperbolic proof of Fermat's last theorem. This is not a matter of merely changing the rules (for example, using a ruler as a measuring device instead of a straightedge). It is more significant than that. Instead, it's a matter of changing whole definitions. And, regardless, it is logically inconsistent to *reject* a

hyperbolic method of squaring the circle and *accept* a hyperbolic method of proving FLT.

There you have it. It is very naughty to base a proof on hyperbolic geometry. Marilyn Vos Savant has expanded the *Parade* column into a book entitled *The World's Most Famous Math Problem (The proof of Fermat's Last Theorem and other mathematical mysteries)*, published by St. Martin's Press (New York, 1993). It is reviewed by Nigel Boston and Andrew Granville in the *American Mathematical Monthly* 102 (1995) 470–473.

15. A quick (?) proof of irrationality

The problem is from a 1996 school textbook; the solution appears in the teacher's edition. Here is a challenge for the reader: can you prove that the decimal expansion of $5^{1/2}$ is nonrepeating without first proving it is irrational?

Problem.   Explain why $4^{1/2}$ is rational while $5^{1/2}$ is irrational.

Solution. $4^{1/2} = 2$ which is rational. $5^{1/2}$, in its decimal form, does not terminate or repeat and therefore cannot be written as an integer over an integer. ♡

Contributed by Richard Askey of the University of Wisconsin in Madison.

16. A rational combination of two transcendentals

Theorem. π^e *is rational.*

Proof.   Observe first that, if r is rational, then $\log_\pi r$ is nonrational. For otherwise, if $\log_\pi r = s$, a rational, then the equation $\pi^s = r$ contradicts the transcendence of π.

   Now suppose, if possible, that $\pi^e \notin \mathbf{Q}$. Then, for all $r \in \mathbf{Q}$, $\pi^e \neq r$, whence $e = \log_\pi \pi^e \neq \log_\pi r$ because the logarithm is a one-to-one function. Therefore, by the initial observation, $e \neq k$ for $k \notin \mathbf{Q}$, which is to say that e is rational. But this is a contradiction. Thus π^e is rational. ♡

Contributed by Christian Counts when he was a sophomore at Loyola Marymount University in Los Angeles, CA.

17. How the factorial works

Norton Starr of Amherst College in Massachusetts notes that, in the book *Go Figure* (Contemporary Books, 1998), Clint Brookhart shows how one can compute $248.3e^{0.0076(60)}$ with a scientific calculator that lacks a e^x key but does have inverse and natural log keys. Is there such a calculator?

More interestingly, on pages 34 and 35, the author explains "how the n factorial works." This quantity occurs "throughout mathematical formulas and expressions, particularly in many types of series (the sum of a usually infinite sequence of numbers). ... Because sums in these series increase rapidly, it is useful to be able to approximate when dealing with large values of n." The tool for this, of course, is Stirling's approximation formula, quoted as

$$n! = \left(\frac{n}{e}\right)^n (2\pi n)^{\frac{1}{2}}.$$

"Let's see," writes the author, "how well Stirling's formula works when $n!$ grows exponentially". He then goes on to calculate $12! (= 4.7569 \times 10^8)$ and $20! (= 2.42278 \times 10^{18})$, and concludes with the comment:

Finally, let's compare the two factorials we computed:

$$\frac{20!}{12!} = \frac{(2.422787 \times 10^{18})}{(4.7569 \times 10^8)} = 5.1 \times 10^9.$$

The summation does grow exponentially!

Dollars and sense

Stuart Mills, Louisiana State University, Shreveport, LA 71115-2399

Suppose that you make $10,000 a year. Your boss offers you a choice. You can have

(1) a $1000 raise (not a bonus) at the end of each year, or

(2) a $300 raise at the end of each six months.

Which should you take?

The question has been around for awhile. It appeared in the March 15, 1992 column *Ask Marilyn*, written by Marilyn Vos Savant [2], and in the July 27, 1983 issue of *The Chronicle of Higher Education* [1], a publication read by many university faculty and administrators.

There has been much debate about which is the better choice. Part of the difficulty is to understand exactly what is offered. Consider these alternatives: Would you choose

(3) a $1000 per year raise at the end of each year, or

(4) a $300 per year raise at the end of each six months?

What would be your choice if offered

(5) a $1000 per year raise at the end of each year, or

(6) a $300 per six months raise at the end of each six months?

Or what about

(7) a $1000 per month raise at the end of each year, or

(8) a $300 per month raise at the end of each six months?

The choice posed in each case is between two sequences of raises. When the raises, which are rates, are clearly stated as in (3) through (8), one can easily choose the sequence of rates which will produce the better income. For the alternative forms given above, the choices are (3), (6) and (7). A change of units is often helpful; for example, one could replace "$300 per six months" with "$600 per year".

What about the original question? The proposed raises, missing the defining unit of time, are stated only as dollar amounts rather than as rates. Let T_1 and T_2 represent the missing units of time. Then the sequence of raises looks something like this:

(1) $1000 per T_1 raise at the end of each year, or

(2) $300 per T_2 raise at the end of each six months.

Most people, upon encountering the choice, subconsciously assign the value "a year" to T_1 and T_2. Respondents are led to this unconscious assignment by the use of a yearly basis in defining the $10,000 annual salary and in the timing of the first sequence of raises on an annual basis.

However, both the editor of *The Chronicle* and Marilyn Vos Savant give a "solution" based on the assignment:

$$T_1 = \text{a year} \qquad T_2 = \text{six months}$$

With these values, (2) is the better choice.

When respondents question the "solution" or seek understanding, their error is explained to them this way:

T_1 must be a year because the timing of the sequence is annual, and T_2 must be six months because the timing of that sequence is semi-annual.

This is the reasoning that is used. However, the language is not so illuminating, for otherwise the ambiguity would be exposed.

Some have argued that these raises are, in fact, well defined, the missing time unit being implied and, therefore, known. For there to be a meaningful solution, the implied or understood values of T_1 and T_2 must be ascertainable, unique and without dispute. Such a case may be made for the assignment $T_1 =$ a year, but an implied value of T_2 is very much in dispute. Most respondents make the assignment $T_2 =$ a year, while the "solution" makes the assignment $T_2 =$ six months. Respondents come to "understand" the latter assignment as "correct" only after much enlightenment. The need to explain proves that there is no clear implied value for T_2, confirming that $T_2 =$ six months is an arbitrary assigned value.

The wording in *The Chronicle* is more precise:

> If your employer offered to raise your salary $300 every six months or $1000 a year, which would you choose?

The *Chronicle* article cites its source of the question as Professor Hiram S. Bleeker, professor of physics at the State University of New York College at Cortland. I do not know if he originated the conundrum, but he adds a significant confounding feature. *The Chronicle* says:

> because of a compounding effect, he (Bleeker) notes, a $300 raise every six months would provide progressively larger increments each year than annual raises of $1000.

The introduction of the totally irrelevant concept of compounding is a brilliant stroke of obfuscation on the part of Professor Bleeker, or else he never had a clue as to what was going on.

References

1. *The Chronicle of Higher Education,* (a) "In Box," p. 15, July 27, 1983; (b) "Letters to the editor," September 14, 1983.

2. Marilyn Vos Savant, *Ask Marilyn*, Parade, March 15, 1992.

Comments

The problem discussed by Mills goes back farther than he indicates. Henry O. Pollak of Summit, NJ, discovered it in the Pelican edition of W.W. Sawyer's *Mathematician's delight* (page 58 of the 1943 edition; pages 68–69 of the 1964 edition). But David Singmaster of South Bank University in London, England, tracked it back even further:

> My earliest record of it is in the 3rd edition of W.W. Rouse Ball's *Mathematical Recreations and Essays*, 1896, pp. 26–27. Ball includes it among several "questions which I have often propounded in past years. ... Two clerks are engaged, one at a salary commencing at the rate of (say) £100 a year with a yearly rise of £20, the other at salary commencing at the same rate (£100 a year) with

a half-yearly rise of £5, in each case payments being made half-yearly". As Mills notes, the phrasing of these problems is often ambiguous, but I think Ball has been reasonably clear.

The problem probably arose from a straightforward exercise, but I know only one example of this — in *The Tutorial Arithmetic* of 1902 (p. 425, no. 16 in my 6th printing of the 2nd edition.) "A Clerk enters upon work at a yearly salary of £80, rising £15 a year every three years. What difference would have been made in his total income for 10 years if he had received an annual increase of £5 a year?"

From then, the problem has been repeated regularly — I have references to 1904, 1905, 1907, 1914, 1917, 1924, 1931, 1932, 1936, 1943 and then I don't bother with later references. In all of these, the interpretation of a £5 rise each six months is as a £10 rise in the yearly salary, except in Sam Loyd's *Cyclopedia* of 1914 (pp. 312 & 381), where he interprets it as £5 per year, so it amounts to only £2.50 each six months.

One of Singmaster's references is to R. Ripley, *Believe It or Not: Book 2* (Simon & Schuster, 1931). On page 123, the reader is invited to marvel that a raise of 1 every day is better than a raise of 35 every week.

Chapter 2

ALGEBRA AND TRIGONOMETRY

I. Do you know how to split the atom?

The April 3, 1994 issue of the *Washington Post* recounted how a sports celebrity failed to answer the following questions on a high school equivalency test:

1. If the equation for a circle is $x^2 + y^2 = 34$, what is the radius of the circle?

2. If $6 - 50 = x + 20$, what is x?

3. If $2x$ plus $3x$ plus $5x = 180$, what is x?

Bert Sugar, the publisher of *Boxing Illustrated,* was not surprised at the failure. He opined that anyone who could answer the math questions "could probably qualify as a nuclear scientist." The reporter's reaction to this view was not recorded.

Contributed by Milt Eisner of Mount Vernon College in Washington, DC.

2. The number of tickets

The following problem is from page 52 of the first edition of *Intermediate algebra* by K. Elayn Martin-Gay (Prentice-Hall, Englewood Cliffs, NJ, 1993); the solution is provided by a student.

Problem. Eight hundred tickets for a play were sold for $2000. If the adult tickets cost $4 each and student tickets cost $2 each, find how many of each kind of ticket was sold.

Solution. Let x tickets be sold. Then $4x + 2x + 800 = 2000$. So $6x = 1200$ or $x = 200$. $\heartsuit$.

In fact, there *were* 200 adult tickets sold.

Contributed by Robert W. Vallin of Slippery Rock University in Penn-sylvania.

3. A superficial volume problem

The following problem appears on page 112 of the text, *College algebra: a problem solving approach,* by Walter Fleming, Dale Varberg & Herbert Kasube (4th ed., Pentice-Hall, 1992):

Problem. A square piece of cardboard was used to construct a tray by cutting 2-inch squares out of each of the four corners and turning up the flaps. Find the size of the original square if the resultant tray has a volume of 128 square inches.

One cannot have a volume of 128 *square* inches. However, it turns out whether the box has a volume of 128 cubic inches or a surface area of 128 square inches the answer is the same.

Contributed by Randall K. Campbell-Wright of the University of Tampa in Florida.

4. The end justifies the means

One hazard (from the examiner's point of view) of a multiple-choice contest is that students may obtain the right answer for the wrong reason. Here is question 19 on the 1988 American High School Mathematics Examination. The pleasure of the examiners that so many students obtained the correct answer must have been dampened by the report of some proctors that certain candidates solved the problem as indicated below.

Question. Simplify

$$\frac{bx(a^2x^2 + 2a^2y^2 + b^2y^2) + ay(a^2x^2 + 2b^2x^2 + b^2y^2)}{bx + ay}.$$

Solution.

$$\frac{bx(a^2x^2 + 2a^2y^2 + b^2y^2) + ay(a^2x^2 + 2b^2x^2 + b^2y^2)}{bx + ay}$$

$$= \frac{bx(ax + by)^2 + ay(ax + by)^2}{bx + ay}$$

$$= \frac{(bx + ay)(ax + by)^2}{bx + ay}$$
$$= (ax + by)^2. \heartsuit$$

Communicated by Leo Schneider of John Carroll University in Cleveland, OH.

5. How to solve a quadratic equation

Problem. Solve the equation $2 - x - x^2 = 0$.

Solution. The equation can be rewritten $4 = 6 - x - x^2 = (x + 3)(2 - x)$. Setting $x + 3 = 4$ and $2 - x = 1$ yields $x = 1$, while setting $x + 3 = 1$ and $2 - x = 4$ yields $x = -2$. Both solutions check out. $\heartsuit$

A student in Florida, asked to solve $x^4 - 3x^2 + 2 = 0$ reduced the equation to $x^2(x^2 - 3) = -2$ and found $x^2 = 2$ (with $x^2 - 3 = -1$) and $x^2 - 3 = -2$ (with $x^2 = 1$), all correct answers. As pointed out in the solution of Problem 298 in *Crux Mathematicorum* 4 (1978) 167–170, the process can be applied quite generally. It was reported in *American Mathematical Monthly* 74 (1967) 856, that in 1965, the third Wisconsin High School Mathematical Talent Search posed a problem based on this method of solving the equation, and then asked whether one could find a quadratic equation $x^2 + bx + c = 0$ with integer roots that could not be written as $(x - d)(-e - x) = f$ ($f \neq 0$) where $d + f$ and $-(e + f)$ are roots of the original equation. In Round 25 of the International Mathematical Talent Search, published in *Mathematics and Informatics Quarterly* 7 (1997) 101, it was asserted that each quadratic equation $(x - p)(x - q) = 0$ allowed for constants a, b, c with $c \neq 0$ for which $(x - a)(b - x) = c$ was equivalent to the original equation and also the faulty reasoning "either $x - a$ or $b - x$ must equal c" yielded the correct answers $x = p$ or $x = q$. Students were then asked to determine such constants for the equation $(x - 19)(x - 97) = 0$.

Comment on the Florida student from Lewis C. Heckroth of Okaloosa-Walton Community College in Niceville, FL.

6. A new method for solving a cubic

Here is a compressed version of the solution to the cubic equation $x^3 - 7x - 6 = 0$ presented by a candidate in the 1994 Euclid (University of Waterloo) high school mathematics competition.

$$x(x^2 - 7) = 6$$

$$x^2 - 9 = x^2 - 7 - 2 = \frac{6}{x} - 2$$

$$x(x - 3)(x + 3) = 6 - 2x$$

$$(x^2 + 3x)(x - 3) + 2(x - 3) = 0$$

$$(x + 1)(x + 2)(x - 3) = (x^2 + 3x + 2)(x - 3) = 0$$

$$x = -1, -2, 3. \diamond$$

More generally, the equation $x^3 - px = q$ can be rendered as $x(x^2 - p - z) + z(x - q/z) = 0$, where z is chosen to make $x^2 - p - z$ divisible by $x - q/z$. Thus, we need to solve the "adjoint" equation $z^3 + pz^2 = q^2$. Are there nice examples that this approach actually yields a significant simplification?

7. An old method for solving a cubic

An issue of the *Physics and Mathematics Digest,* published in Singapore, contains this technique for solving a cubic equation. Suppose that we wish to solve

$$x^3 + px + q = 0.$$

Set $x = y \cos \theta$. Then we obtain

$$\cos^3 \theta + \frac{p}{y^2} \cos \theta + \frac{q}{y^3} = 0.$$

But

$$\cos^3 \theta - \frac{3}{4} \cos \theta - \frac{1}{4} \cos 3\theta = 0.$$

Hence

$$\frac{p}{y^2} = -\frac{3}{4} \quad \text{and} \quad \frac{q}{y^3} = -\frac{1}{4} \cos 3\theta,$$

so that $y^2 = -(4p/3)$ and $\cos 3\theta = -(4q/y^3)$ determine y and θ.

For example, trying this on $x^3 - 3x - 1 = 0$ yields $y = 2$, $\cos 3\theta = 1/2$ or $3\theta \equiv \pi/3, 5\pi/3 \pmod{2\pi}$. The solutions are

$$x = 2 \cos \frac{\pi}{9} = 1.87938\ldots, \quad 2 \cos \frac{5\pi}{9} = -0.34729\ldots,$$

$$2 \cos \frac{7\pi}{9} = -1.53209\ldots.$$

($y = -2$ leads to the same trio.) Is this method on the level? ♣

The justification can be given along these lines. The substitution $x = y \cos \theta$ introduces an extra degree of freedom. Accordingly, we can select y so that $p/y^2 = -3/4$; thus, θ must satisfy the conditional equation $\cos^3 \theta + (-3/4) \cos \theta + \frac{q}{y^3} = 0$. Comparing this with the identity for $\cos 3\theta$ yields $q/y^3 = (-1/4) \cos 3\theta$.

Irving K. Feinstein of the University of Illinois at Chicago observes that this is a "stripped-down version of the method of François Viète (1540–1603)", who used a combination of the substitutions $y = mx$ and $y = \cos \theta$ to first choose m and then determine θ. He notes that m will be real only if all three roots are real, and suggests that this method is not as popular as others because of the weakness where the roots are not all real. For an account of how Viète employed this approach to solve a challenge problem, consult page 59 of *Trigonometric delights* by Eli Maor (Princeton University Press, Princeton, NJ, 1998).

8. An exponential equation

Problem. Solve for x:

$$e^{x-2} + e^{x+8} = e^{4-x} + e^{3x+2}.$$

Solution. The "identity" $e^a + e^b = e^{ab}$ yields

$$(x-2)(x+8) = (4-x)(3x+2),$$

which reduces to $0 = x^2 - x - 6 = (x-3)(x+2)$. Hence $x = -2$ or $x = 3$. Both these solutions check out. ♡

The equation $b^{f(x)} + b^{g(x)} = b^{u(x)} + b^{v(x)}$ will have solutions $x = r$ and $x = s$ if we can choose functions such that $f(r) = u(r)$, $g(r) = v(r)$, $f(s) = v(s)$, $g(s) = u(s)$. This is true regardless of the base b. The numbers r and s will also satisfy the equation $f(x)g(x) = u(x)v(x)$. To perpetrate the flimflam, just arrange that all solutions of the last equation satisfy the earlier equation, for example by making both sides of the last equation polynomials of degree 2.

9. Logarithms distribute over sums

Problem. Solve for x:

$$e^{2x} - e^x - 2 = 0.$$

Solution.

$$\log e^{2x} - \log e^x - \log 2 = 0 \implies 2x - x - \log 2 = 0 \implies x = \log 2.\heartsuit$$

This and the next item were contributed by Robert H. Thompson of Washburn University of Topeka, KS as examples of students' work.

10. The multiplication rules for logarithms

Problem.   Solve for x:

$$2\log(x - 2) = \log(x - 3) + \log x.$$

Solution.

$$\log(2x - 4) = 2\log(x - 2) = \log(x - 3) + \log x = \log(x^2 - 3x)$$
$$\implies 2x - 4 = x^2 - 3x$$
$$\implies 0 = x^2 - 5x + 4 = (x - 4)(x - 1).$$

Hence $x = 4$ ($x = 1$ is inadmissible). $\heartsuit$

11. A lack of technical unanimity

It is normally held that the equation

$$\log(x^2 - 10) = \log(10x - 40) - 1$$

(logarithms to base 10) has no real number solutions. The software MICRO-CALC and many graphing calculators agree, but DERIVE and the TI-92, among others, show graphically and symbolically, that $x = 3$ and $x = -2$ are valid solutions. Who is correct?

Contributed by Carlton A. Lane of Dale Mobry Campus of Hillsborough Community College in Tampa, FL.

12. A straightforward cancellation

Problem.   Solve for x and simplify

$$\binom{n}{n - r} x = \binom{n}{r + 1}.$$

Solution.

$$x = \binom{n}{r+1} \Big/ \binom{n}{n-r} = \binom{n}{r+1}\binom{n-r}{n} = \frac{n-r}{r+1}. \spadesuit$$

We note that

$$\frac{\binom{a}{c}}{\binom{a}{b}} = \frac{a/c}{a/b} = \frac{b}{c}$$

if and only if $b = c$ or $a + 1 = b + c$.

Contributed by Sally Ringland of Clarion University of Pennsylvania.

I3. An application of the Cauchy-Schwarz Inequality

A letter from a student contained the following sentences: "I came across an interesting question, about which I consulted some of my math teachers and several other enthusiastic math students. They however are as puzzled as I am, and we are unable to solve it." The question and solution submitted with the letter was the following:

Problem. If a, b, c are real numbers such that $a^2 + b^2 + c^2 = 1$, find the set of possible values for $ab + bc + ca$.

Solution. By the Cauchy-Schwarz Inequality,

$$1 = (a^2 + b^2 + c^2)(b^2 + c^2 + a^2) \geq (ab + bc + ca)^2,$$

whence $-1 \leq ab + bc + ca \leq 1$. $\heartsuit$

The student continues: "The lower bound is wrong! Since $(a+b+c)^2 \geq 0$, it follows that $1 + 2(ab + bc + ca) \geq 0$, so $ab + bc + ca \geq -\frac{1}{2}$. The lower bound should have been further restricted to $-\frac{1}{2}$. But shouldn't the Cauchy-Schwarz inequality always be right? Where is the error?" What would you tell the student?

I4. Surprising symmetry

David Wells, in his book *You are a mathematician* (John Wiley & Sons, 1995, p. 88), makes the interesting observation that the nonsymmetric condition $a = b+c$ leads to the symmetric result $a^4 + b^4 + c^4 = 2b^2c^2 + 2c^2a^2 + 2a^2b^2$. Does this mean that $a = b + c$ is equivalent to $b = c + a$ and $c = a + b$? $\clubsuit$

The nature of the symmetry becomes evident when we factor:

$$(2b^2c^2 + 2c^2a^2 + 2a^2b^2) - (a^4 + b^4 + c^4)$$

$$= (a + b + c)(a + b - c)(b + c - a)(c + a - b).$$

If any of the variables is equal to the sum of the other two, the symmetric equation obtains. On the other hand, the symmetric equation leads either to $a + b + c = 0$ or one of three symmetrically related conclusions. We can understand what is happening geometrically by interpreting the factored difference as 16 times the square of the area of a triangle with sides a, b, c.

I5. Factoring homogeneous polynomials

Proposition. *Let f be a homogeneous polynomial in several variables with a factorization $f = gh$. Then the polynomials g and h are homogeneous.*

Proof. Write $g = u + v$, where u is homogeneous and deg $v <$ deg $u =$ deg g. Then $gh = uh + vh$ and deg $vh <$ deg $uh =$ deg f. Since f is homogeneous, we must have $vh = 0$, whence $v = 0$ and g is homogeneous. Similarly h is homogeneous. ♠

The erroneous argument appears on page 276 in the first printing of *Polynomials* by E.J. Barbeau (Springer-Verlag, New York, 1989). Since h may not be homogeneous, the terms of lower degree in uh may cancel the terms of vh. The difficulty can be evaded by writing also $h = p + q$ where p is homogeneous and deg $q <$ deg p.

Contributed by John Webb and Graeme West of the University of Cape Town in South Africa, who published a note in Pythagoras (the journal for the Association for Mathematics Education of South Africa), no. 33 (April, 1994), page 4.

I6. Polynomial detection

The author of a paper, not accepted by a journal, wished to establish that, for each positive integer k, the sum of the kth powers of the first n integers is a polynomial in n. He noted that the sum $1^k + 2^k + \cdots + n^k$ just consists of powers of numbers (which belong to the set of polynomials). The sum of polynomials being a polynomial, the result follows.

I7. The remainder theorem

Theorem. *If a polynomial $f(x)$ is divided by $x - c$, then the remainder is $f(c)$.*

Proof. Divide x into $f(x)$, getting $f(\ )$. Multiply $f(\ )$ times $x - c$ to get $f(x) - f(c)$. Subtracting, we get $f(c)$. Symbolically

$$
\begin{array}{r}
f(\ ) \\
x-c\overline{)f(x)} \\
\underline{f(x)-f(c)} \\
f(c)
\end{array}
$$

♡

Contributed by Richard Laatsch of Miami University in Oxford, OH.

18. The zero polynomial

Two colleagues were examining this proof that only the zero polynomial could take the value zero everywhere.

Proposition.   Suppose that a polynomial $p(t) = a_n t^n + a_{n-1} t^{n-1} + \cdots + a_0$ is identically equal to 0; that is, $p(t) = 0$ for all t. Then $a_n = a_{n-1} = \cdots = a_0 = 0$.

Proof.   Suppose, if possible, that the polynomial has positive degree n. Since $p(0) = p(1) = \cdots = p(n) = 0$, the factor theorem provides that $p(t) = t(t - 1)(t - 2) \cdots (t - n)q(t)$ for some nonzero polynomial $q(t)$. Since the degree of the left side is n and of the right side is at least $n + 1$, we obtain a contradiction. ♠

Omicron: This proof that only the zero polynomial vanishes everywhere is wrong. It appeals to the factor theorem to show that $p(t)$ has positive degree n to get $p(t) = t(t - 1) \cdots (t - n)q(t)$, and uses the disparity of degrees on the two sides to get a contradiction.

Upsilon: So, what's wrong with that?

O: The proof assumes what has to be shown. Denoting the right side by $f(t)$, the prover wants the result that, if $p(t) = f(t)$ everywhere, then $p(t)$ and $f(t)$ must have the same degree. This seems to depend on knowing that corresponding coefficients of $p(t)$ and $f(t)$ agree, or that, because $p(t) - f(t)$ vanishes everywhere, its coefficients are all zero. But this is what has to be established. ♣

U: Hmm. You make a telling point. But we should look more closely. Do you agree that we should distinguish two types of equality between polynomials? There is the weak sense that $p(t) = f(t)$ for each individual value of t. For example, if m is prime, then, as polynomials over a finite field with m elements, t^m equals t in the weak sense. Then there is the strong algebraic sense that $p(t) = f(t)$ because corresponding coefficients agree. If in the proof, $p(t) = f(t)$ in the weak sense, then the objection can be sustained.

O: I suppose that you are trying to tell me that $p(t) = f(t)$ in the strong sense.

U: Sure. Let us see how the equality in the proof is obtained. Because $p(0) = 0$, it is clear that the constant coefficient vanishes and we can write $p(t)$ as t times, say, $p_1(t)$, whose degree is one less than that of $p(t)$. Now divide $p_1(t)$ by $t - 1$. This is effected by continually subtracting from $p_1(t)$ the product of $t - 1$ and a constant multiple of a power of t until there is a constant remainder. Since $p_1(1) = 0$, the remainder vanishes and we have $p_1(t) = (t - 1)p_2(t)$ where the degree of $p_2(t)$ is one less than the degree of $p_1(t)$. We can continue on in this way to get $p(t) = t(t - 1) \cdots (t - n)q(t)$. This equation can be checked by multiplying the terms of the right side and comparing coefficients. However, because of this business of each division lowering the degree by 1, such an equation is untenable.

O: My mind reels at all this complexity. Let's go and have a coffee.

19. An inductive fallacy

Theorem. *Every polynomial f with rational coefficients factors into irreducible polynomials all of the same degree.*

Proof. We proceed by induction on the number of irreducible factors of f. Clearly, if f has only one irreducible factor, then all factors of f have the same degree. Now assume the statement is true for polynomials that factor into n irreducible factors. Suppose that f is a polynomial that factors into $n + 1$ irreducible factors $f_1, f_2, \ldots, f_{n+1}$. Then the product $f_1 f_2 f_3 \cdots f_n$ has n irreducible factors, so $f_1, f_2, \ldots, f_n$ are all of the same degree by the induction hypothesis. Similarly, the product $f_2 f_3 \cdots f_{n+1}$ has n irreducible factors which must be all of the same degree. Therefore f_{n+1} is of the same degree as f_n which is of the same degree as $f_1, f_2, \ldots, f_{n-1}$, so all of the factors f_i are of the same degree. ♡

On comparing this argument with the more familiar argument that every horse has the same color (cf., Chapter 4, Item 1), the reader will perceive that this indeed is not a horse of another color.

Contributed by Adrian Riskin and William Stein of North Arizona University in Flagstaff.

20. On not identifying equations and identities

Richard Askey of the University of Wisconsin in Madison has an extensive knowledge of school texts and does not always approve of what he finds in them. He comments, "It is generally believed that with the end of the New Math, text books were watered down, and this was part of the reason for the need to change our school mathematics curriculum. As an illustration of what was written in 1980, a few years after the New Math was seen to be dead by essentially everyone, consider this example from a precalculus book. The book will not be named since there are surely others which would do equally well to explain what happened. Notice that some of what was said to be part of the New Math is still there, namely formal explanations of things which earlier might not have been defined."

An identity tells you that two expressions are equal for all values of the variable involved. Thus

$$\tan x = \frac{\sin x}{\cos x}$$

is an identity because it is true for every x for which both sides are defined. On the other hand,

$$\tan x = 1$$

is true only for certain values of x, and so this is an equation.

Proving that something is in fact an identity is called verifying the identity. In order to do this we have to work separately with the expressions on either side of the equals sign, and somehow show that they are equal to one another. The problem is that an identity looks very like an equation — both have an equals sign set between two expressions — and so it is very easy to slip into doing things that are O.K. for solving equations but not for verifying identities. For example, in solving an equation we frequently move things from one side of the equation to the other. But when we're verifying an identity, we're trying to prove that the two expressions on either side of the equals sign are identical, and we can't possibly do that if we move things from one side to the other. Again in solving an equation we often multiply both sides by some quantity, but in verifying an identity we can't do that, as it changes the values of the expression that we're trying to prove equal.

This passage is followed by a box with the following advice:

To Verify an Identity: Treat each side separately, and manipulate each as an expression until they are both in the same form. Don't change the value of the expression on either side by
> Moving things from one side to the other
> Adding/subtracting/multiplying/dividing by anything.

This means that you can
> Factor
> Multiply Out
> Add fractions
> Simplify
> Substitute
> Use the basic identities

If in doubt, it is usually helpful to write everything in terms of sines and cosines.

"So what is the author trying to say?" continues Askey. "I wrote and asked, but never got a response. I think I know. When I was in school, the usual instruction was to take one side and change it until you got the other side. No reason was given for this, but when my teacher was asked, he replied in a similar fashion. If you move terms from one side to the other, you have a different identity. I knew this was nonsense, but if the rules were to only work on one side, I was willing to play by those rules. A few years later I figured out why this is the right set of rules. What you are trying to do is to simplify an expression, not to prove an identity. The reason the two sides are given is that students are not experienced enough to know what 'simple' is. Also, what is simple can change from one setting to another.

"Mark Saul, a school teacher in the New York area and the Mathematics Editor of *Quantum* told me that the way he presents identities to students is to give them two lists of expressions, ask the students to pair them and show that those in each pair are equal. That is a nice refinement. Once you have an identity to prove, you should try to do it by working with one side. If you cannot prove the identity this way, then do whatever you have to to prove the identity, as long as you can undo the steps. This includes moving terms from one side to the other, and multiplying both sides of the identity by the same function. After you have proven the identity, you are not done. You need to use the knowledge you have gained to redo the problem in the right way, just working on one side. Thus, the mathematical advice is wrong and so is the teaching advice.

"How is it that the advice was so wrong, both from my trigonometry teacher and the author of this book? My teacher was asked a question to which he did not have a good answer, and he wanted us to do things the way they were done in the book, so found an answer which had satisfied students. Whether he believed it or not, I do not know. I suspect that he never thought about this question, at least not seriously. The author of the current text was probably educated from books like I used, which had the claim about working on one side but no reason why. This was not a satisfactory state of affairs after the introduction of many reasons in the New Math, so something had to be said. Clearly it was felt to be too restrictive to only allow students to work on one side of an identity, so a less restrictive format was proposed. Unfortunately, the reason given for not allowing an even less restricted format was wrong. Again, I do not know if the author would have known better with serious thought, but it is clear that there was no serious thought."

21. A surd equation

Exercise. Solve the equation

$$\sqrt{x} + \sqrt{x - a} = 2$$

where a is a positive real parameter.

Solution. The given equation implies also that

$$\sqrt{x} - \sqrt{x - a} = \frac{a}{\sqrt{x} + \sqrt{x - a}} = \frac{a}{2}.$$

Adding the two equations together, we obtain $2\sqrt{x} = 2 + a/2$, whence $x = (1 + \frac{a}{4})^2$. It is readily checked that $x - a = (1 - \frac{a}{4})^2 > 0$, so that the surd is defined. ♡

However, if $a = 8$, this solution yields $x = 9$, which is strange, since the left side of the equation now exceeds 2. ♣

Observe that solving an equation involves the initial assumption that the equation *is* viable. The given equation requires that $x \geq a$, so that $2 = \sqrt{x} + \sqrt{x - a} \geq \sqrt{a}$ or $2 \geq a/2$ is a necessary condition for consistency. Alternatively, noting that $\sqrt{x} + \sqrt{x - a} \geq \sqrt{x} - \sqrt{x - a}$, we get the requirement $2 \geq a/2$. Thus, when $a \leq 4$, we can solve the equation as indicated. For example, $a = 4$ leads to $x = 4$, and $a = 1$ to $x = 25/16$.

When $a > 4$, the equation has no real solutions. However, we can use the indicated technique to solve $\sqrt{x} - \sqrt{x - a} = 2$ and get $x = (1 + \frac{a}{4})^2$, $x - a = (1 - \frac{a}{4})^2$. In this case, note that $\sqrt{x - a} = a/4 - 1$.

The reader is invited to sketch the graph of the equation $y = \sqrt{x} + \sqrt{x - a}$ when $a < 4$ and when $a > 4$ and see under what circumstances it intersects the line of equation $y = 2$.

22. The disappearing solution

Since there is a one-one relationship between real numbers and their cubes, the following equations in real variables x and y all appear to be equivalent:

$$(x+y)^{1/3} + (x-y)^{1/3} = 1 \tag{1}$$

$$2x + 3(x^2 - y^2)^{1/3} = [(x+y)^{1/3} + (x-y)^{1/3}]^3 = 1 \tag{2}$$

$$x^2 - y^2 = \left(\frac{1-2x}{3}\right)^3 \tag{3}$$

$$y^2 = x^2 - \left(\frac{1-2x}{3}\right)^3. \tag{4}$$

Setting $x = -1$, we obtain from the final equation that $y = 0$. But these values of x and y do not satisfy the first equation. ♣

Leon Gerber of St. John's University in Jamaica, NY provides this brief analysis. Let $a = (x+y)^{1/3}$ and $b = (x-y)^{1/3}$. Then $a+b = 1$, from which it follows from cubing that

$$a^3 + 3ab(a+b) + b^3 = 1.$$

Since $a+b = 1$, we deduce that $a^3 + 3ab + b^3 = 1$. Note that this step is not reversible. Indeed, $a^3 + 3ab + b^3 = 1$ leads to

$$0 = a^3 + 3ab + b^3 - 1 = (a+b-1)(a^2 - ab + b^2 + a + b + 1)$$
$$= \frac{1}{2}(a+b-1)\left[(a+1)^2 + (b+1)^2 + (a-b)^2\right],$$

from which we obtain either that $a+b = 1$ or the spurious solution $(a,b) = (-1,-1)$ which corresponds to $(x,y) = (-1,0)$.

Sung Soo Kim of Hanyang University in Ansan, Korea observes that many students appreciate the difficulty after being told that $(u,v) = (-1,0)$ fails to satisfy $1 = (u+v) + (u-v)$ but does satisfy the equation obtained from cubing:

$$1 = [(u+v) + (u-v)]^3 = (u+v)^3 + 3(u^2 - v^2) + (u-v)^3$$
$$= 2u^3 + 6uv^2 + 3(u^2 - v^2).$$

Despite the apparently real character of the situation, we can obtain a way of relating the solutions to the outside equations by using complex numbers. Let us look again at the first equation in the case that $x = -1$ and $y = 0$. Then a suitable interpretation of the nonreal cubic surds will in fact give a real sum: in the first, interpret $\sqrt[3]{-1}$ as $\frac{1}{2}(1 + i\sqrt{3})$ and in the second, as $\frac{1}{2}(1 - i\sqrt{3})$. Then, the sum of the two terms on the left is indeed 1, as desired. We delve into the situation a little more. It is clear that any solution of (1) satisfies (4), so the issue turns on in what sense the converse is true.

We will get some insight by introducing the variable $t = y^2$ and allowing t to take negative values for a "transitional" phase. The graph of the equation

$$t = x^2 - \frac{1}{27}(1 - 2x)^3 = \frac{1}{27}(x + 1)^2(8x - 1) \qquad (5)$$

is sketched below. Since the derivative of the right side is $(2/9)(4x+1)(x+1)$, we can identify a local maximum on the graph at $(-1, 0)$ and a local minimum at $(-\frac{1}{4}, -\frac{1}{16})$. Observe how the values of x for which $t = 0$ relate to those for which $t \geq 0$; note that, if $t > 0$, then the corresponding real value of x is uniquely determined.

When $t > 0$, the quantities $x + y = x + \sqrt{t}$ and $x - y = x - \sqrt{t}$ are real, and it is straightforward to verify that exactly one of the nine choices of cube roots for this pair will make the left side of (1) real. When y is nonzero, there is a unique value of x for which (4) is satisfied and this pair (x, y) will also satisfy (1).

The case $t = y = 0$ corresponds to more than one solution of (4), since it is now satisfied by $(x, y) = (\frac{1}{8}, 0)$ and $(-1, 0)$. Where u is the real cube

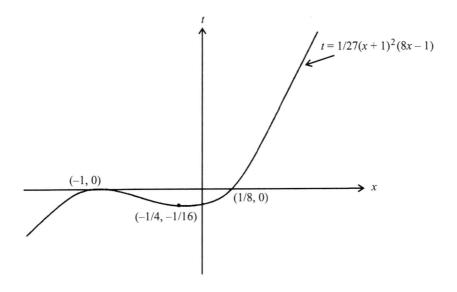

FIGURE 2.22

root of x, the left side of (1) has two possible real values, namely $2u$ and $u(\omega + \omega^2) = -u$, where ω is an imaginary cube root of unity. Exactly one of these is equal to 1. We mediate between the two values of x for which $y = 0$ by looking at the situation when $t < 0$.

Let $t = -s^2 < 0$ for $s > 0$, so that $y = is$. We study the real values of

$$z \equiv (x + is)^{\frac{1}{3}} + (x - is)^{\frac{1}{3}}. \tag{6}$$

Suppose that, for some angle θ with $0 < \theta < 60°$, we have that

$$\cos 3\theta = \frac{x}{r} \qquad \sin 3\theta = \frac{s}{r}$$

where $r = \sqrt{x^2 + s^2}$. Then $\sqrt[3]{x + y} = \sqrt[3]{x + is}$ has three possible values: $r^{1/3}(\cos\theta + i\sin\theta)$, $r^{1/3}(\cos(\theta + 120°) + i\sin(\theta + 120°))$ and $r^{1/3}(\cos(\theta + 240°) + i\sin(\theta + 240°))$; the other cube root $\sqrt[3]{x - is}$ similarly has three possible values, the complex conjugates of these. Selecting the cube roots so that the value of (6) is real makes it equal to one of:

$$2r^{1/3}\cos\theta, \quad 2r^{1/3}\cos(\theta + 120°), \quad 2r^{1/3}\cos(\theta + 240°).$$

Suppose that the value of (6) is $z = 2r^{1/3}\cos\theta$. Then

$$z^3 - 2x = 8r\cos^3\theta - 2r\cos 3\theta = 2r[\cos 3\theta + 3\cos\theta] - 2r\cos 3\theta = 6r\cos\theta$$

so that

$$x^2 - \left(\frac{z^3 - 2x}{3z}\right)^3 = r^2\cos^2 3\theta - r^2 = -r^2\sin^2 3\theta = -s^2 = t.$$

A similar check can be made when z is replaced by the two other possibilities. Thus, each substitution for the cube roots in (7) leads to an equation

$$y^2 = t = x^2 - \left(\frac{z^3 - 2x}{3z}\right)^3. \tag{7}$$

At most one of these equations can have $z = 1$.

In general, we see that equations (6) and (7) are related in the way that (1) and (4) are. Since (7) is equivalent to

$$\frac{t}{z^6} = \left(\frac{x}{z^3}\right)^2 - \left(\frac{1 - 2(x/z^3)}{3}\right)^3,$$

the correspondence $(x, t, z) \leftrightarrow (x/z^3, t/z^6, 1)$ relates solutions of (5) and (7), so that the graph of (7) is like that of (5), with $t = 0$ corresponding to $x = z^3/8$ and $x = -z^3$.

Observe that the two values of x that correspond to $t = 0$ in (5) are connected by an open interval $(-1, 1/8)$ for which the related values of t

are negative. Suppose we let x vary along the closure of this interval from right to left in such a way that t varies continuously with x according to (5). Thus, t ranges over the interval $[-1/16, 0]$, assuming its minimum value when $x = -1/4$.

What happens to

$$\sqrt[3]{x + \sqrt{t}} + \sqrt[3]{x - \sqrt{t}} \tag{8}$$

the left side of (1)? Let s, r and θ have the meanings assigned above, and, when $(x, y) = (1/8, 0)$, assign to each of the cube roots in (8) the positive value $1/2$. As x decreases from $1/8$, t, s, r and θ vary continuously with (5) holding. Because of our assignment at $x = 1/8$ and the continuity, we use the value $r^{1/3}(\cos\theta + i\sin\theta)$ for the cube root of $x + \sqrt{t}$. As x decreases to -1, $\cos 3\theta = x/r$ varies from 1 to -1, so that θ goes from $0°$ to $60°$. Through all of this, (6) and (8) maintain the value $2r^{1/3}\cos\theta = 1$. For example, when $x = 0$, then $t = -1/27$, $s = r = 1/3\sqrt{3}$, $\theta = 30°$ and (8) has the value

$$2r^{1/3}\cos 30° = \left(\frac{2}{\sqrt{3}}\right)\left(\frac{\sqrt{3}}{2}\right) = 1.$$

When $x = -1$, (8) is equal to 1, but with $\sqrt[3]{x + y} = \sqrt[3]{-1}$ assigned the nonreal value

$$r^{1/3}(\cos\theta + i\sin\theta) = (\cos 60° + i\sin 60°) = \frac{1}{2}(1 + i\sqrt{3}).$$

In a similar way, we can start with a different assignment of the cube roots at the right end of the interval $(-1, 1/8)$ and wind up with a real assignment at the left end.

Notice that (7) is satisfied by both $(x, t, z) = (-1, 0, 1)$ and $(-1, 0, -2)$. For the first solution, the determination $r^{1/3}(\cos\theta + i\sin\theta)$ with $0 \le \theta \le 60°$ for $\sqrt[3]{x + y}$ makes the left side of (6) equal to 1 when $\theta = 60°$. For the second solution, the determination $r^{1/3}(\cos(\theta + 120°) + i\sin(\theta + 120°))$ with $0 \le \theta \le 60°$ for $\sqrt[3]{x + y}$ makes the value of (8) equal to -2. In this case, the choice of cube root of $x + y$ is real when $\theta = 60°$ and nonreal when $\theta = 0°$.

23. Solving an inequality

Problem.   Solve for x the inequality $|x| + |x - 1| < 2$.

Solution.   Combining the terms of the left side, we find that the inequality is equivalent to $|2x - 1| < 2$, whereupon $-1/2 < x < 3/2$. ♡

The answer is, in fact, correct. Readers are invited to determine conditions on a, b, c, with $a \le b$, for which the solution sets of the inequalities

$|x - a| + |x - b| < c$ and $|2x - (a + b)| < c$ are the same. This fluke was analyzed in a class paper by Angie Wittig while a student in a course of Andrew Balas at the University of Wisconsin-Eau Claire.

24. An appearance of finite geometric sequences

Problem.   Let u be a real number. Determine all finite sequences $\{a_1, a_2, \ldots, a_n\}$ for which

$$a_m = u(a_{m+1} + a_{m+2} + \cdots + a_n) \qquad (m = 1, 2, \ldots, n - 1).$$

Solution.   For each m,

$$a_m = u a_{m+1} + u(a_{m+2} + \cdots + a_n) = u a_{m+1} + a_{m+1},$$

whence $a_{m+1} = (1 + u)^{-1} a_m$. Hence, $\{a_1, a_2, \ldots, a_m\}$ is a geometric sequence with common ratio $1/(u + 1)$. Of course, this does not work with $u = -1$, in which case there is no sequence with the required property. ♠

The first equation of the solution is not valid when $m = n - 1$, as $a_{n-1} = u a_n$. If $a_n = a$, then the finite sequence must be

$$(u(1 + u)^{n-2} a, u(1 + u)^{n-3} a, \ldots, u(1 + u)^2 a, u(1 + u)a, ua, a)$$

so that $a_k = u(1+u)^{n-k-1} a$ for $1 \leq k \leq n - 1$. In particular, when $u = -1$, the sequence is $(0, 0, \ldots, 0, 0, -a, a)$.

25. Glide-reflecting the sine curve

Here is a student howler with a mishap at almost every step.

Problem.   Prove that $\sin(x + \pi) = -\sin x$.

Solution.

$$\sin(x + \pi) = \sin x + \sin \pi = \sin x - 1 = (-1) \sin x = -\sin x. \qquad \heartsuit$$

Contributed by Norman Rice of Queen's University in Kingston, ON.

26. A trigonometric identity

In the course of a solution on the 1998 Descartes Competition administered by the Canadian Mathematics Competition at the University of Waterloo in

Ontario, a candidate stated without justification that

$$\sin^2 A - \sin^2 B = \sin(A + B)\sin(A - B).$$

Aha! thought the marker, clearly a case of rendering the left side as $(\sin A + \sin B)(\sin A - \sin B)$ and then taking the factor "sin" out of each term. But the result is legitimate: the left side can be written as

$$\frac{1}{2}(\cos 2B - \cos 2A) = \frac{1}{2}[\cos(\overline{A + B} - \overline{A - B}) - \cos(\overline{A + B} + \overline{A - B})]$$

en route to obtaining the right side.

However, an unwary student might be tempted to infer from the fact that $\sin(A + B)\sin(A - B) = (\sin A + \sin B)(\sin A - \sin B)$ that $\sin(A \pm B) = \sin A \pm \sin B$. J. Messenger in the *Mathematical Gazette* 58 (1974) 215 (note 309) reported on a question on a 1970 Joint Matriculation Board (UK) A Level paper. Candidates were asked to determine $\sin 105°$. One student wrote

$$\sin 105° = \sin 45° + \sin 60° = \frac{1}{\sqrt{2}} + \frac{\sqrt{3}}{2} = \frac{1 + \sqrt{3}}{2\sqrt{2}}.$$

Commented upon by J.D. Aczel of the University of Waterloo in Ontario and R.P. Boas in Seattle, WA.

27. Floored by an Olympiad problem

The following problem appeared on the first paper in the 1992 Australian Mathematical Olympiad. The solution is due to a student using it as a practice problem.

Problem. Let n be a positive integer. Determine how many real numbers x with $1 \le x < n$ satisfy

$$x^3 - \lfloor x^3 \rfloor = (x - \lfloor x \rfloor)^3.$$

Solution. Let $x = a + b$ where $a \in \{1, 2, \ldots, n - 1\}$ and $0 \le b < 1$, Then $x^3 = a^3 + 3a^2 b + 3ab^2 + b^3$ so that, since $0 \le b^3 < 1$, we must have one of the three possibilities (i) $\lfloor x^3 \rfloor = a^3$, (ii) $\lfloor x^3 \rfloor = a^3 + \lfloor 3a^2 b \rfloor$ and (iii) $\lfloor x^3 \rfloor = a^3 + \lfloor 3a^2 b + 3ab^2 \rfloor$. Plugging these into the given equation yields, in case (i), that $3ab(a + b) = 0$ whence $b = 0$ and we have $n - 1$ possibilities. In cases (ii) and (iii), we are led respectively to the equations

$$0 = 3ab(a + b) - \lfloor 3a^2 b \rfloor,$$

$$0 = 3ab(a + b) - \lfloor 3ab(a + b) \rfloor.$$

In both cases, the equation cannot occur since $3ab(a+b)$ is not an integer and so the right side is strictly positive. Hence, in all, there are $n-1$ solutions to the equation. ♠

To focus ideas, specialize to $n = 2$. Then on each interval $[k^{1/3}, (k+1)^{1/3})$, $(1 \leq k \leq 7)$, the function $x^3 - \lfloor x^3 \rfloor$ increases from 0 towards 1, while $(x - \lfloor x \rfloor)^3$ increases from 0 to 1 on $[1, 2)$. From the graphs of the two functions, it is straightforward to see that the equation is satisfied for 6 values of x.

With $x = 1 + b$, the given equation is equivalent to

$$3b(b+1) = \lfloor 3b + 3b^2 + b^3 \rfloor$$

where $0 \leq b < 1$. The function $3b(b+1)$ can assume the integer values 0, 1, 2, 3, 4, 5 over the domain $[0, 1)$ of values of b. If b is selected to make $3b(b+1)$ an integer, then, since

$$3b(b+1) \leq 3b + 3b^2 + b^3 < 3b(b+1) + 1,$$

we have that $\lfloor 3b + 3b^2 + b^3 \rfloor = 3b(b+1)$ as desired. In the above "solution," we see that case (ii) yields no possibilities, but that in case (iii), $3ab(a + b)$ can indeed be an integer. The correct answer to the problem is $n^3 - n$.

28. A New Identity for the Ceiling Function

An anonymous contributor noted that, although Donald Knuth popularized the notation for the ceiling function, in his series *The Art of Computer Programming* he did not prove a host of identities about it as he did for the binomial coefficients and Fibonacci numbers. Recall that $\lceil x \rceil = m$ if and only if m is an integer for which $m - 1 < x \leq m$. The contributor provided this result and assured us that, if methods like this gain wide application, then every conceivable identity for the ceiling and floor functions will make its way into the literature.

Theorem.   *For any real number x, $\lceil 3x \rceil + \lceil x \rceil = 2\lceil 2x \rceil$.*

Proof.   Consider the equation $\lceil x + y \rceil + \lceil x \rceil = 2\lceil y \rceil$. We will show that $y = 2x$ is a solution for any real x.

Because either $\lceil x + y \rceil = \lceil x \rceil + \lceil y \rceil$ or $\lceil x + y \rceil = \lceil x \rceil + \lceil y \rceil - 1$, we will break $\lceil x + y \rceil + \lceil x \rceil = 2\lceil y \rceil$ into the two equations

$$\lceil x \rceil + \lceil y \rceil + \lceil x \rceil = 2\lceil y \rceil \tag{1}$$

and

$$\lceil x \rceil + \lceil y \rceil + \lceil x \rceil - 1 = 2\lceil y \rceil \qquad (2)$$

Since $\lceil x \rceil$ is constant on an interval of the form $(m-1, m]$ with m an integer, we look for integer solutions, $x = m$ and $y = n$, of (1) and (2), and extend the solutions to the square $\{(x, y) : m - 1 < x \le m, n - 1 < x \le n\}$.

If $x = m$ and $y = n$ are integers, then equation (1) reduces to $2m = n$ and equation (2) reduces to $2m - 1 = n$. Thus, (x, y) solves $\lceil x + y \rceil + \lceil x \rceil = 2\lceil y \rceil$ if $m - 1 < x \le m$ and $2m - 1 < y \le 2m$ or if $m - 1 < x \le m$ and $2m - 2 < y \le 2m - 1$.

Now consider the line $y = 2x$ so $\lceil y \rceil = \lceil 2x \rceil$, and let $m = \lceil x \rceil$. If $\lceil x \rceil - x < 1/2$, then $2m - 1 < 2x \le 2m$, so $\lceil 2x \rceil = 2m$; if $\lceil x \rceil - x \ge 1/2$, then $2m - 2 < 2x \le 2m - 1$, so $\lceil 2x \rceil = 2m - 1$. This shows that if $y = 2x$, either (1) or (2) is satisfied. Thus, we have shown that for any real number x, $\lceil 3x \rceil + \lceil x \rceil = 2\lceil 2x \rceil$. ♠

Note that $y = 2x$ and (1) or (2) would imply $\lceil 3x \rceil + \lceil x \rceil = 2\lceil 2x \rceil$ in the presence of $\lceil x + y \rceil = \lceil x \rceil + \lceil y \rceil$ or $\lceil x + y \rceil = \lceil x \rceil + \lceil y \rceil - 1$ *respectively*.

Let $\lceil x \rceil = m$. It is true that if $\lceil x \rceil - x < \frac{1}{2}$, then $\lceil 2x \rceil = 2m$ and $\lceil x \rceil + \lceil 2x \rceil + \lceil x \rceil = 4m = 2\lceil 2x \rceil$. This along with $\lceil 3x \rceil = \lceil x \rceil + \lceil 2x \rceil$ would certainly imply $\lceil 3x \rceil + \lceil x \rceil = 2\lceil 2x \rceil$. But the latter property is not a necessary consequence of $\lceil x \rceil - x < \frac{1}{2}$. For, if $\frac{1}{3} \le \lceil x \rceil - x < \frac{1}{2}$, then $\lceil 2x \rceil = 2m$, $\lceil 3x \rceil = 3m - 1$ and $\lceil 3x \rceil = \lceil x \rceil + \lceil 2x \rceil - 1$.

Similarly, $\lceil x \rceil - x \ge \frac{1}{2}$ implies $\lceil 2x \rceil = 2m - 1$ and $\lceil x \rceil + \lceil 2x \rceil + \lceil x \rceil - 1 = 4m - 2 = 2\lceil 2x \rceil$. This along with $\lceil 3x \rceil = \lceil x \rceil + \lceil 2x \rceil - 1$ would imply $\lceil 3x \rceil + \lceil x \rceil = 2\lceil 2x \rceil$. But when $m - \frac{2}{3} < x \le m - \frac{1}{2}$, $\lceil 2x \rceil = 2m - 1$ and $\lceil 3x \rceil = 3m - 1$, so that $\lceil 3x \rceil \ne \lceil x \rceil + \lceil 2x \rceil$.

Counterexamples to the theorem are $x = 2/5$ and $x = 3/5$.

Chapter 3

GEOMETRY

I. The impossibility of angle bisection

In a typical introductory course in abstract algebra, after you have proven the impossibility of trisecting an arbitrary angle using just straightedge and compasses, you sum up the argument as follows: "We have just shown that $\cos 20°$ is not constructible, and so we cannot construct a $20°$ angle either; thus we cannot trisect a $60°$ angle, and so we cannot trisect an arbitrary angle."

You can often create some consternation by continuing: "Now the fact that we cannot construct a $20°$ angle also shows that we cannot *bisect* a $40°$ angle and so you cannot bisect an arbitrary angle with compasses and straightedge." ♣

Since an angle bisection *is* possible with straightedge and compasses, all that has been shown is that an angle of $40°$ is not so constructible. If a $40°$ angle was given, it would have had to have been determined by some measuring device. A $60°$ angle is constructible, so if a trisection were possible, we would be able to obtain a $60°$ angle and then trisect it to obtain a $20°$ angle.

Contributed by Eric Chandler of Randolph-Macon Woman's College in Lynchburg, VA.

2. Trisecting an angle with ruler and compasses

Construction. Let the angle to be trisected be BAC. With center A and respective radii of two, three and four units, draw arcs PU, QV and RW to intersect the arms of the angle. Determine D, E, F and G, the respective midpoints of arcs PU, RW, RE and EW. Let M and N be the respective

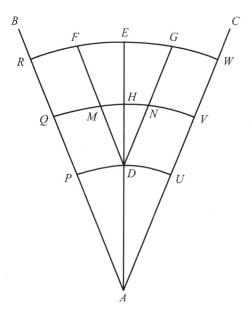

FIGURE 3.2.1

intersections of the segments FD and GD with the arc QV. Then the rays MA and NA yield the desired trisection of angle BAC.

First proof. Let H be the midpoint of arc QV. Consider the "triangles" DMH and DFE, one side of each being a circular arc. Since the arcs RW and QV are parallel, $\angle DFE = \angle DMH$ and $\angle DEF = \angle DHM$, so that triangle DMH is similar to triangle DFE. Since $2DH = DE$, it follows that arc $MN = 2$ arc $MH =$ arc FE. Now, arc $QV = (3/4)$ arc $RW = 3$ arc FE and arc $QM =$ arc NV. Therefore, the arc QV is trisected by M and N, and so the construction is valid.

Second proof. Since arc $RW = 2$ arc PU, arc $PD =$ arc RF. Therefore, FD is parallel to RP, and so arc $QM =$ arc RF. Similarly, arc $NV =$ arc $GW =$ arc RF. Since arc $QV = 3$ arc RF, QV is trisected by M and N. ♠

　　Where do these proofs run into trouble? In the first, the assertion that the two "triangles" with common vertex are similar is confounded by the fact that the curved sides are not arcs of circles with their centers at the common vertex. The angle between a curve and a line at a point is defined to be the angle between the tangent to the curve and the line. We can see that the

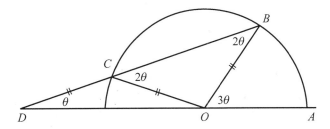

FIGURE 3.2.2

tangents to the circular arcs at F and M are not parallel, as the line joining F to A would intersect the arc $QMNV$ in a point where the tangent to the arc is parallel to the tangent at F. In the second, because the two equal arcs between the two lines do not have the same radius, we cannot conclude that the lines are parallel. One way to approach the situation is through transformations. Two lines are parallel if and only if there is a translation of the plane that takes one to the other. Note that the chord RE is parallel to the chord PD as they are related by a dilation with center A. The translation that takes P to D takes R to the midpoint of the chord RE. Since FA but not FD passes through this midpoint, it is seen that RP and FD are not parallel.

If we allow the straightedge to be marked, then the trisection problem becomes solvable. Figure 3.2.2 illustrates the method.

AOB is the angle to be trisected, with A and B on an arbitrary circle with center O. Mark the length OA on the ruler; draw a line through B to cut AO produced at D, using the mark to determine C on the circle so that $DC = OA$. This method is classical and described on page 136 of David M. Burton, *The history of mathematics: an introduction* (Allyn and Bacon, Newton, MA, 1985).

Construction contributed by Tom Cunningham of Brockville, ON. For other examples of trisections, consult The Trisectors by Underwood Dudley, published by the Mathematical Association of America.

3. A luney way to square a circle

The mathematics director of the Athenian Academy sighed as he rolled open the scroll. Another circle-squaring construction, submitted by a chap identified only as H! Well, he would look it over before passing it on to a graduate student to draft a reply. For convenience, we will summarize the work in modern terms.

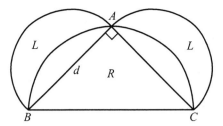

FIGURE 3.3.1

Suppose we are given a circle of diameter d. The problem is to construct, using straightedge and compasses, a square of area equal to that of a circle. As in Figure 3.3.1, construct an isosceles right triangle ABC whose equal sides have length d. Let the hypotenuse BC be the diameter of a semi-circle passing through A, and let semi-circles also be constructed on diameters AB and AC. The area of the larger semi-circle is equal to twice that of each of the smaller, and it is not hard to argue that the sum of the areas of the two lunes (labeled L) is equal to the area of the triangle (labeled R).

Now construct a trapezoid $DEFG$ which is the upper part of a regular hexagon of side d. Thus $DG = 2DE = 2EF = 2FG = 2d$. The area of the semi-circle with diameter DG is four times the area S of the semi-circle of diameter d constructed on each of the sides DE, EF, FG as diameter. It can be seen that the area S plus the area of the three lunes (L) is equal to the area of the trapezoid (T).

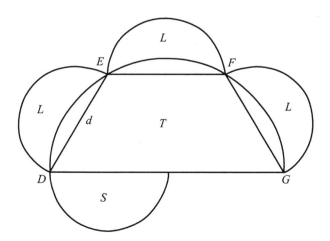

FIGURE 3.3.2

Symbolically, we have $R = 2L$ and $T = 3L + S$. Hence the area of the given circle is $2S = 2T - 6L = 2T - 3R$. Thus, we have been able to construct rectilinear figures some linear combination of which will yield the area of the circle. It is known that one can construct with straightedge and compasses a square whose side is equal to $2T - 3R$. ♣

For further details on this incident, consult *A history of Greek mathematics* (Volume 1) by Thomas Heath (Oxford, 1921: pp. 183–186). The difficulty in the construction is that the shorter arcs with the congruent chords in the two figures arise from circles with different radii, so the lunes for one figure are not congruent to the lunes for the second figure.

4. The Steiner-Lehmus Theorem

The Steiner-Lehmus theorem is simply stated, but notoriously difficult to prove. The theorem was sent to Jacob Steiner in 1840 by C.L. Lehmus. The complicated proof by Steiner prompted a long search for a shorter one. The following "proof" from a textbook was reproduced in *Crux Mathematicorum* (Eureka) 2 (1976) 92–93, 174–175. Will you be drawn in? A correct proof can be found in *Geometry revisited* by H.S.M. Coxeter and S.L. Greitzer, Mathematical Association of America, 1967, p. 14. There is also a proof by J.V. Malesević in *Mathematics Magazine* 43 (1970) 101–102, and a criticism of this proof as not being direct by M. Lewin in *Mathematics Magazine* 47 (1974) 87–89. For additional results inspired by this theorem, consult *Mathematical Gazette* 57 (1973) 336–339.

Proposition. *If in a triangle two angle bisectors are equal, then the triangle is isosceles.*

Proof. Let BAC be the triangle and AN, CM the two equal bisectors, with N and M on BC and AB respectively. Suppose the perpendicular bisectors of AN and CM meet at O. The circle with center O passes through A, M, N, C. Angles MAN and MCN, subtended by MN are equal. Hence, the angles BAC and BCA are equal, and the result follows. ♠

What is wrong with the proof? It has the perpendicular bisectors of the two equal angle bisectors AN and CM meeting at O. However, there is no guarantee that O is equidistant from AN and CM. Therefore, we cannot claim without further justification that the circle with center O that contains A and N is the same as the circle with center O that contains C and M.

5. A geometry problem

The October, 1998 discussion draft of *Principles and Standards for School Mathematics* (NCTM) reproduces on page 293, a geometry example said to be from the book *Secondary Mathematics 5B* by Ho Juan Beng (Singapore Ministry of Education, Pan Pacific Publishing, 2nd ed., 1995). The example is headed by this homily:

> Students in grades 9–12 should be able to answer questions and prove theorems about geometric situations even when the diagrams that depict them are somewhat complex. Teachers might modify a task such as the one in figure 7.12 to assess students' inclinations to find relationships in a more open-ended way. For example, teachers might provide only the figure and ask students to find a pair of congruent triangles or a pair of similar triangles, or to list other relationships that hold for the figure. Students can then be asked to justify their claims on the basis of earlier theorems and facts.

Figure 7.12 is the following, with the accompanying questions:

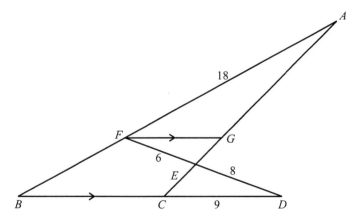

Questions. In the figure, $AF = BD$, $FG = FB$, and FG is parallel to BD.
 (a) Name the triangle that is congruent to $\triangle AFG$,
 (b) Prove that $\triangle AFE$ is similar to $\triangle DCE$, and
 (c) Given that $AF = 18$ cm, $FE = 6$ cm, $CD = 9$ cm, and $DE = 8$ cm, calculate CE and EG.

Now let us see whether this problem is suitably complex. Let x be the common length of FB and FG. From the similar triangles EFG and EDC, we have that $x : 6 = 9 : 8$ whence $x = 27/4$. Applying Menelaus' Theorem to the triangle ABC and transversal FED, we find that $BF \cdot AE \cdot CD = FA \cdot EC \cdot BD$ whence $AE : EC = 16 : 3$.

The similarity of triangles AFE and DCE is equivalent to the respective lengths of AE and EC actually being *equal* to 16 cm and 3 cm. But if this is the case, then because of the congruence of triangles AFG and DBF (SAS), we find that AG has the same length as DF, namely 14 cm. But then the respective lengths of EG and CG are 2 cm and 5 cm. Thus, $x : 18 = BF : FA = CG : GA = 5 : 14$, so that $x = 45/7$.

However, for sure, triangles AFG and ABC are similar, which leads to

$$\frac{18}{x} = \frac{18 + x}{9}$$

and $x = 9(\sqrt{3} - 1)$.

This example was discovered by Kim Mackey of Valdez High School in Alaska, who posted it in the *mathed* discussion group on the net.

6. A case of irregularity

Problem. Given a unit square $IJKL$ with center at O, construct the two medians from each vertex to the nonadjacent sides as in Figure 3.6.1. Find the area of the octagon with vertices at the intersections A, B, C, D, E, F, G and H.

First solution. Note that Area(octagon) = 8Area(triangle OAB). To determine the area of triangle OAB, we note that A is the midpoint of OM, and thus segment OA has length $\overline{OA} = 1/4$. Also $\overline{OB} = 1/4$, since tri-

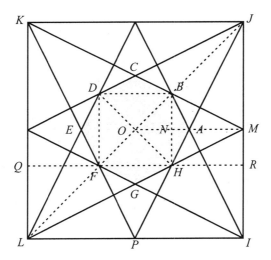

FIGURE 3.6.1

angle OAB is isosceles. Triangle ONB is a right isosceles triangle, hence $\overline{BN} = \overline{OB}/\sqrt{2} = 1/(4\sqrt{2})$. Thus, the area of triangle OAB is equal to (base)(height)/2 = $\overline{OABN}/2 = 1/(32\sqrt{2})$. Finally, the area of the octagon is 8 times this, namely $1/(4\sqrt{2})$.

Second solution.

Area(octagon) = Area(square $FHBD$) + 4Area(triangle BHA).

We first note that JP is the median from vertex J to side LI of triangle JLI, and thus $\overline{FH} = \overline{HR}$. By a similar argument, $\overline{FH} = \overline{QF}$, and since $\overline{QF} + \overline{FH} + \overline{HR} = 1$, they must each be 1/3. Likewise, $\overline{BH} = \overline{DB} = \overline{FD} = \overline{NM} = 1/3$. Since $\overline{NM} = 1/3$ and $\overline{AM} = 1/4$, we have $\overline{NA} = \frac{1}{3} - \frac{1}{4} = \frac{1}{12}$. Thus, the area of the octagon is equal to $(\frac{1}{3})^2 + 4 \cdot \frac{1}{2} \cdot \frac{1}{3} \cdot \frac{1}{12} = \frac{1}{6}$. ♠

Which of these two solutions is correct? The first solution holds that $OA = OB$, which would occur if and only if the inner octagon is regular. However, taking $\alpha = \angle MLI$, we find that $\alpha = \arctan(1/2)$ and the angles of the inner octagon are alternately the unequal angles $\pi - 2\alpha$ and $(\pi/2) + 2\alpha$. A more general setting for the problem is given as problem 343 in *Five Hundred Mathematical Challenges* by E.J. Barbeau, M.S. Klamkin and W.O. Moser (MAA, 1995). This problem is featured on a cover of *Mathematics Magazine*, for the issue containing the article "Dörrie tiles and related miniatures" by Edward Kitchen (*MM* 67 (1994) 128–139). The square can be partitioned into four kites, four darts and the central octagon by two diamonds as in Figure 3.6.2. The central octagon, each kite and the four darts taken together all have area 1/6.

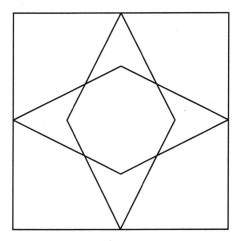

FIGURE 3.6.2

Two solutions contributed by Herb Bailey and Al Schmidt of Rose-Hulman Institute of Technology in Terre Haute, IN. Mathematics Magazine reference supplied by Steven R. Conrad of Manhasset, NY. Observation about kites and darts from James E. Kessler of Vermont Technical College in Randolph Center.

7. A counterexample to Morley's Theorem

Suppose that ABC is an arbitrary triangle and that the trisectors of each of the vertex angles are drawn. Then, according to Morley's Theorem, the intersection points of adjacent pairs of the trisectors are the vertices of an equilateral triangle. However, as Figure 3.6 illustrates, this is not how the geometry software *Cabri* sees the matter.

Contributed by William Watkins of California State University at Northridge.

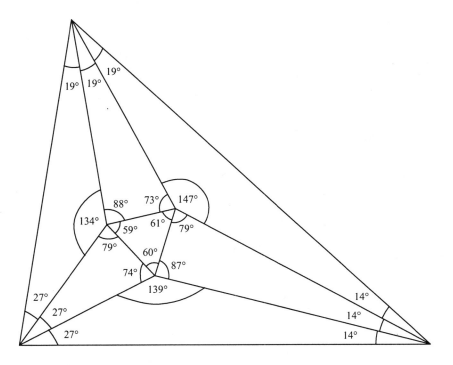

FIGURE **3.7**

8. Going for the stars

Rick Mabry of the Louisiana State University in Shreveport, LA has an item on his web page http://www.lsus.edu/sc/math/rmabry/folding /fivefold.htm in which he learned how to create a perfect (?) five-pointed star by folding a standard $8\frac{1}{2}'' \times 11''$ piece of paper and wielding a pair of scissors. It was something that he picked up from his wife, a middle-school teacher who had learned it at a conference.

> With the sheet oriented with its long side horizontal, fold the left side to the right and crease; the crease joins the midpoints B and C of the longer sides of the sheet. The crease BC is now the left side; the left bottom corner is C. By making a second fold, we can locate the midpoint M of the top of the folded sheet. Fold the bottom left corner C of the folded sheet up to M and crease. Let Q and R be the intersection of this crease with, respectively, the left edge (BC) and right edge of the folded sheet. Make another fold to bisect the angle MQR and fold the triangular flap BQM back about QM. Now make a snip with your scissors at a suitable angle across the folded sheet, unfold and a five-point star is revealed. Is it really perfect?

Anyone wishing to explore other methods for creating stars should consult the article: Steven I. Dutch, Folding $n-$pointed stars and snowflakes. *Mathematics Teacher* 87 (1994) 630–637 and the brief account of the technique of Betsy Ross in *Mathematics Teacher* 88 (1995) 720.

9. Identifying the angle

Problem. ABC is an isosceles triangle with $AB = AC$. The point D is selected on the side AB so that $\angle DCB = 15°$ and $BC = \sqrt{6}AD$. Determine the degree measure of $\angle BAC$.

Solution.   Let $AB = AC = 1$ and let $\angle DCA = \alpha$, where $0 < \alpha < 75°$. Then $BC = 2\cos(15° + \alpha)$. The Sine Law applied to triangle ADC yields

$$\frac{1}{\sin(30° + \alpha)} = \frac{CD}{\sin(150° - 2\alpha)}$$

whence

$$CD = \frac{\sin(150° - 2\alpha)}{\sin(30° + \alpha)}.$$

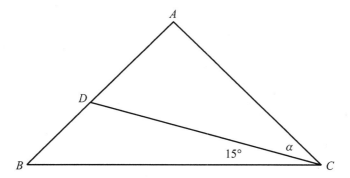

FIGURE 3.9.1

Applying the Sine Law to triangle DBC yields

$$\frac{2\cos(15° + \alpha)}{\sin(30° + \alpha)} = \frac{BC}{\sin(150° - \alpha)} = \frac{CD}{\sin(15° + \alpha)}$$
$$= \frac{\sin(150° - 2\alpha)}{\sin(15° + \alpha)\sin(30° + \alpha)}.$$

Hence

$$\sin(30° + 2\alpha) = 2\cos(15° + \alpha)\sin(15° + \alpha) = \sin(150° - 2\alpha)$$

so that $30° + 2\alpha = 150° - 2\alpha$ with the result that $\alpha = 30°$. Hence $\angle BAC = 150° - 2\alpha = 90°$. ♠

This checks out: $BC = \sqrt{2}$ and $AD = 1/\sqrt{3}$. However, the argument does not use the information about the ratio of BC and AD, and so applies whenever $\angle DCB = 15°$. The equation $\sin(30° + 2\alpha) = \sin(150° - 2\alpha)$ has two possible consequences. Either $30° + 2\alpha$ and $150° - 2\alpha$ are equal or they sum to $180°$. But the latter is always true, so it appears that the argument makes no progress towards the desired result.

One way to obtain the result is to apply the Sine Law to triangles DBC and ADC to obtain

$$\frac{v}{u} = \frac{\sqrt{6}\sin(15° + \alpha)}{\sin(30° + \alpha)} = \frac{\sin(30° + 2\alpha)}{\sin\alpha}$$

where v and u are the respective lengths of CD and AD. This simplifies to

$$\sqrt{6}\sin\alpha = 2\cos(15° + \alpha)\sin(30° + \alpha) = \sin(45° + 2\alpha) + \sin 15°.$$

Letting $\theta = \alpha + 15°$, we find that

$$(\sqrt{6} - 2\cos\theta)\sin\theta\cos 15° = (\sqrt{6} + 2\cos\theta)\cos\theta\sin 15°.$$

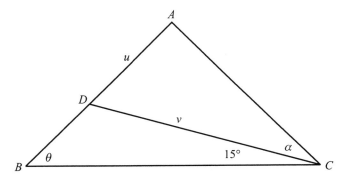

FIGURE 3.9.2

Now

$$\left(\frac{\sqrt{6} - 2\cos\theta}{\sqrt{6} + 2\cos\theta}\right)\tan\theta$$

is an increasing function of θ for $0 < \theta < 90°$, and, when $\theta = 45°$, it assumes the value

$$\frac{\sqrt{6} - \sqrt{2}}{\sqrt{6} + \sqrt{2}} = 2 - \sqrt{3} = \tan 15°,$$

so that the equation is satisfied only for the acute angle $\theta = 45°$. This yields $\alpha = 30°$ and $\angle BAC = 90°$.

Contributed by K. R. S. Sastry of Bangalore, India.

10. The speeder's delight

The situation. Poised 50 feet to the side of the road, a speed trap awaits 120 feet beyond the crest of a hill. In a flash, the officer pulls me over and accuses me of going 65 in a 60 mile-per-hour zone. I say that there has been a math error: Even though the radar gun registered 65, I was going only 60. As everyone knows, the Pythagorean theorem enables us to compute the third side of the triangle in Figure 3.10.1. So if the radar gun indicated 65 (along its line of sight), then by similar triangles my actual speed (in the direction of the road) was 60, as in Figure 3.10.2. ♣

This situation is given to his class by Carl E. Crockett of the United States Air Force Academy in Colorado. He asks his students to prepare a response for the prosecution to use in court. There are three possible responses:

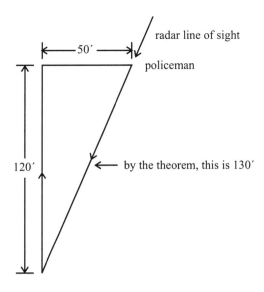

radar line of sight

policeman

50′

120′

by the theorem, this is 130′

FIGURE 3.10.1

1. The velocity of the car can be expressed as the result of two components, one in the direction of the line of sight of the radar equipment and one in the direction perpendicular to this. These two components, equal to the speed of the car multiplied by the cosine of a suitable angle, will both not exceed the speed in magnitude. Thus, we should have drawn the triangle in Figure 3.10.2, from which we find that the speed s of the car satisfies $(12/13)s = 65$ or $s = 70.4$.

2. While there is a speed triangle with the dimensions in Figure 3.10.2, it does not correspond to the given situation. According to this triangle, the defendant could explain a reading of 65 mph by showing a motion of 60 mph in the direction of the road *and* a motion of 25 mph perpendicular to the road. Since there is no motion perpendicular to the road, the argument is not valid.

3. Pythagoras' theorem works for distances. Recognize that (Figure 3.10.3) x and s (distances) are both changing as time passes: $x = x(t)$ and $s = s(t)$. The rate of change of $x(t)$ is the speed of the car and is denoted $x'(t)$. Similarly, $s'(t)$ is the speed indicated by the radar. The theorem tells us that $x^2 + y^2 = s^2$, but it does not say that $x' = kx$, $y' = ky$ and $s' = ks$. The latter was (improperly) implicitly assumed when the defendant used proportions from the distance triangle to (incorrectly) determine the speed.

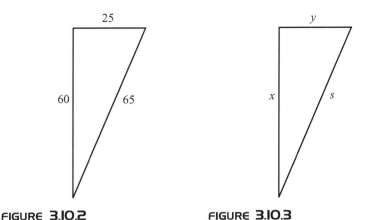

FIGURE 3.10.2   FIGURE 3.10.3

The correct relationship between $x'(t)$ and $s'(t)$ is determined by implicitly differentiating the equation $x^2 + y^2 = s^2$ to get $2xx' + 2yy' = 2ss'$. In this situation, $y' = 0$, so $x' = (s/x)s' = (13/12)65 = 70.4$. The actual speed is always greater than or equal to the radar gun's indicated speed, since s is always greater than or equal to x. In this case, the actual speed was over 70 mph.

II. A solution to problem 480

Problem. Let A', B' and C' denote the feet of the altitudes in the triangle ABC lying on the respective sides BC, CA and AB, respectively. Show that $AC' = BA' = CB'$ implies that ABC is an equilateral triangle.

Solution. Let $k = AC' = BA' = CB'$ and let $u = CA'$, $v = AB'$, $w = BC'$. By the Law of Cosines,

$$(k + u)^2 = (k + v)^2 + (k + w)^2 - 2(k + v)(k + w) \cos A$$

whence

$$(1 - 2\cos A)k^2 + 2((v+w)(1 - \cos A) - u)k + (v^2 + w^2 - u^2 - 2vw \cos A) = 0.$$

Equating coefficients to zero yields in particular that $1 - 2\cos A = 0$ or $\theta = 60°$. ♠

This is problem 480 in the *College Mathematics Journal* 23 (1992) 248; 24 (1993) 275–276. The solution treats the quadratic in k as an identity, which it is not. Consequently, its vanishing indicates the mutual dependence of the variables k, u, v, w and A, and so we cannot conclude that the coefficients vanish. As it happens, however, when $A = 60°$ and $u = v = w$, all the coefficients do vanish.

A correct solution provided by Milton P. Eisner proceeds as follows. Since the center is in the interior of the triangle, all angles must be acute. Assume, say, that $a \leq b \leq c$ so that $A \leq B \leq C < 90°$. Since $AC' = b \cos A$, $BA' = c \cos B$, $CB' = a \cos C$ are all equal,

$$\cos A = \frac{a}{b} \cos C \leq \cos C$$

so that $A \geq C$. Hence $A = C$ and the triangle is equilateral.

Contributed by Dale K. Hathaway of Olivet Nazarene University, Kankokee, IL.

12. Tangency by double roots

In the days before calculus, one way to check the tangency of two curves with algebraic equations $f(x, y) = 0$ and $g(x, y) = 0$ at a common point (a, b) was to eliminate one of the variables from the system of two equations and to check whether the resulting equation in the other variable had a double root corresponding to the common point. As a simple example, $y = x^2$ and $y = 2x - 1$ represent curves tangent at $(1, 1)$ because $x^2 = 2x - 1$ has a double root at $x = 1$. However, this process does have its pitfalls.

Problem. Find all values of k for which the curves with equations

$$y = x^2 + 3 \qquad \text{and} \qquad \frac{x^2}{4} + \frac{y^2}{k} = 1$$

are tangent.

Solution. Eliminating x yields the equation

$$4y^2 + ky - 7k = 0 \tag{1}$$

for the ordinates of the intersection points of the two curves. If the curves are to be tangent, the quadratic equation should have a double root, so that its discriminant $k^2 + 112k$ vanishes. Since $k = 0$ is not admissible, k must be -112. ♡

With the aid of a sketch, it is not hard to see that $k = 9$ also works. Why is it not turned up by this argument? ♣

Observe that the elimination of y leads to the equation

$$4x^4 + (24 + k)x^2 + (36 - 4k) = 0 , \tag{2}$$

a quadratic in x^2 with discriminant $k(k + 112)$. When we solve for the intersection points of the two curves, each root of (1) corresponds to a pair of roots of (2) with opposite signs. Now, let us see why tangency is not always accompanied by a double root.

First, suppose that k is positive, so that the curves are a parabola with its axis along the y-axis and an ellipse centered at the origin. For $k < 9$, x^2 must be negative for each root of (2), so that while there is a corresponding real value of y satisfying (1), the values of x are pure imaginary. Thus, the curves do not intersect. When $k > 9$, (2) has two roots of opposite sign whose squares are positive and two whose squares are negative. The first two correspond to two points of intersection that have the same y-value. Thus, one of the roots of (1) is a positive value of y giving the ordinate of *both* intersection points, and the other is negative (since their product $-7k/4$ is negative) and corresponds to no real point of intersection. As k appoaches 9, the two intersection points coalesce into one. There is no doubling up of the roots of (1), but of course (2) has $x = 0$ as a double root for $k = 9$.

Next, suppose that k is negative. To have any real solutions at all, we must have $k \leq -112$. Let $k < -112$. The curves, a parabola and an hyperbola, have four intersection points, two with positive abscissae and separate ordinates and their reflected images in the y-axis. As k approaches -112, the two positive absissae and the two ordinates coalesce, and we find that at $k = -112$, (1) has $y = 14$ as a double root and (2) has $x = \sqrt{11}$ and $x = -\sqrt{11}$ both as double zeros. In this case, the double root criterion turns out to be valid.

13. A puzzling graph

Consider the general equation of the non-slant conic section, namely,

$$ax^2 + by^2 + cx + dy + e = 0.$$

Of course, a and b cannot both be zero. The graph of such an equation is determined by four (coplanar) points. For example, $3x^2 - 2y^2 - 3x + 4y = 0$ is the equation of the only non-slant conic containing the points $(1, 0)$, $(0, 2)$, $(2, 3)$, $(0, 0)$.

Yet two distinct conics can intersect in four points, as in Figure 3.13. For example, there are many conics that contain the four points $(1, 1)$, $(1, -1)$, $(-1, 1)$, $(-1, -1)$; any equation of the form $ax^2 + by^2 - a - b = 0$ will do. It is quickly seen that a similar situation arises for the general cubic, namely,

$$ax^3 + by^3 + cx^2y + dxy^2 + ex^2 + fy^2 + gxy + hx + iy + j = 0$$

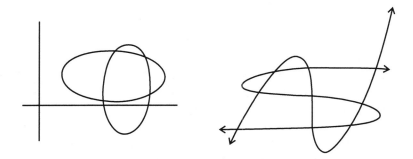

FIGURE 3.13

and the *nine* points determining it.

With respect to the conic situation, compare Problem 1720 in *Crux Mathematicorum*:

> *The osculating circle at point P (not a vertex) of a conic intersects the conic in one other point Q. Find a simple construction for Q, given the conic itself, its axes and the tangent at P.*

A solution by Dan Pedoe with comments by Chris Fisher appears in *Crux Mathematicorum* 19 (1993) 54–55. The circle in question can be considered as intersecting the conic in three coincident points at P, so we have a degenerate case of two conics, one a circle, intersecting in "four" points.

If two non-slant conics intersect in four points, then infinitely many conics, including a circle, contain the four points. Suppose the two distinct conics have the equations

$$ax^2 + by^2 + 1 = 0 \qquad (ab \neq 0, \ a \neq b)$$

and

$$p(x+h)^2 + q(y+k)^2 + 1 = 0 \qquad (pq \neq 0, \ p \neq q).$$

Then, for any λ, the four points of intersection satisfy the equations

$$0 = (ax^2 + by^2 + 1) + \lambda[p(x+h)^2 + q(y+k)^2 + 1]$$
$$= (a + \lambda p)(x+m)^2 + (b + \lambda q)(y+n)^2 + r,$$

where m, n, and r are uniquely determined in terms of a, b, p, q, h, k, λ (by completing the square). When $a + \lambda p = b + \lambda q$, this represents a circle. If A, B, C, D are the common points of intersection, then the chords AB and CD are equally inclined to the axes of the conics; this is proved in the *Crux* solution cited above. Furthermore, we can conclude that, if the unique circle through any three points does not contain a fourth point, then there cannot be a non-slant conic through the four points, although there would be an infinity of other conics.

While a cubic curve is uniquely determined by nine points in "general position," it is possible for two, and thus infinitely many cubics to contain special sets of nine points. Furthermore, any cubic passing through any eight of the points common to two cubics must pass through the ninth common point also. A good reference on these topics is Dan Pedoe, *Geometry: a contemporary approach* (Dover, 1988). Section 79 (pages 340–344) treats conics through four points while Section 93 (pages 417–423) treats cubics.

The original observation in this item is due to Richard L. Francis of Southeast Missouri State University in Cape Girardeau, MO. Additional comments are from Dan Pedoe of Minneapolis, MN. Larry Zimmerman of Brooklyn, NY and Scott Hochwald of the University of North Florida in Jacksonville have prepared a paper on the Euler-Cramer paradox of distinct cubics intersecting in nine points. The history includes Maclaurin, Euler, Cramer, Lamé, Gergonne, Plücker, and Charlotte Scott. As an application of the theorem cited above about cubics passing through eight points, they give the result: Let M and N be arbitrary points on the respective sides AB and CD of parallelogram $ABCD$. Let DM and AN intersect in P, and BN and CM intersect in Q. Then the line PQ bisects the area of the parallelogram.

It suffices to show that PQ contains R, the intersection of the diagonals AC and BD. The union of three lines can be considered as a cubic curve, whose equation is the product of three linear equations. The cubic curves

$$AN \cup MC \cup DB \qquad \text{and} \qquad DM \cup NB \cup AC$$

intersect in A, M, B, C, N, D, P, Q and R. The cubic curve $DC \cup AB \cup PQ$ contains A, M, B, C, N, D, P and Q, and so must contain R. Since R lies on neither AB nor CD, it must lie on PQ. A similar argument proves Pascal's and Pappus' theorems; see problem 93.5 on page 423 of Pedoe's book.

14. The wilting lines

Figure 3.14 is a *Mathematica* printout resulting from a request to graph $y = mx$ for various values of m by asking for the level curves of $z = y/x$. The moral of the story is that you *still* cannot divide by 0.

This was contributed by Randall K. Campbell-Wright of the University of Tampa in Forida. M. Douglas McIlroy of AT&T Bell Laboratories in Murray Hill, NJ claimed to have "found the truth about the wilting lines." Figure 2C on page 826 of the article by E.M. Buhl, K. Malasy, and P. Somogyi,

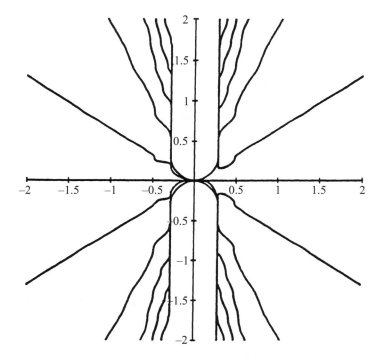

FIGURE 3.14

"Diverse Sources of Hippocampal Unitary Postsynaptic Potentials and the
Number of Synaptic Release Sites" in *Nature* 368 (April 28, 1994) 823–828)
"clearly shows they are hippocampal synapses. For want of anything better,
Mathematica has apparently adapted the brain's native way to divide by zero."

15. The height of a trapezoid

On a linear algebra examination, students were asked to find the area of
an isosceles trapezoid defined by extending the parallelogram determined by
vectors $\mathbf{u}$ and $\mathbf{v}$ making an angle of θ.

In computing the height $h = \|\mathbf{v}\| \sin \theta$ of the trapezoid, a student presented
the following howler:

$$h = \|\mathbf{v}\| - \|Proj_{\mathbf{v}}\mathbf{u}\|$$

$$= \|\mathbf{v}\| - \|\mathbf{v}\| \cos \theta$$

$$= \|\mathbf{v}\| \cdot |1 - \cos \theta|$$

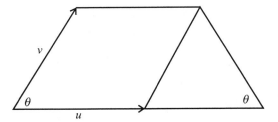

FIGURE 3.15

$$= \|\mathbf{v}\| \sqrt{(1 - \cos \theta)^2}$$
$$= \|\mathbf{v}\| \sin \theta. \ \Diamond$$

Contributed by Dale R. Buske of St. Cloud State University in Minnesota.

16. Forces with a given resultant

Here are alternative solutions to a statics problem, one of which delivers more possibilities than the other.

Problem.   A force $\mathbf{R}$ of magnitude 200 N (Newtons) is the resultant of two forces $\mathbf{F}$ and $\mathbf{G}$ for which $2|\mathbf{F}| = 3|\mathbf{G}|$ and the angle between the resultant and $\mathbf{G}$ is twice the angle between the resultant and $\mathbf{F}$. Determine the magnitudes of $\mathbf{F}$ and $\mathbf{G}$.

Both solutions use the parallelogram representation of the vectors as illustrated, where $3u = |\mathbf{F}|$, $2u = |\mathbf{G}|$ and $v = |\mathbf{R}| = 200$. From the Law of Sines, we have that

$$\frac{\sin 2\theta}{3} = \frac{\sin \theta}{2}$$

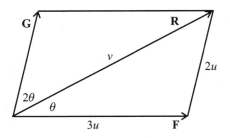

FIGURE 3.16

whence $\cos\theta = 3/4$, $\cos 2\theta = 1/8$ and $\cos 3\theta = 4\cos^3\theta - 3\cos\theta = -9/16$. From here, there are two ways to proceed:

(i) From the Law of Cosines, we find that

$$v^2 = 4u^2 + 9u^2 - 12u^2\cos(180° - 3\theta) = 13u^2 + 12u^2\cos 3\theta = \frac{25u^2}{4}$$

so that $200 = \frac{5}{2}u$, $u = 80$, $|\mathbf{F}| = 240$ and $|\mathbf{G}| = 160$.

(ii) From the Law of Cosines, we find that

$$4u^2 = 9u^2 + v^2 - 6uv\cos\theta = 9u^2 + 200^2 - 900u$$

so that

$$0 = 5(u^2 - 180u + 8000) = 5(u - 80)(u - 100).$$

Hence, $u = 80$, $|\mathbf{F}| = 240$, $|\mathbf{G}| = 160$ or $u = 100$, $|\mathbf{F}| = 300$, $|\mathbf{G}| = 200$.

Why does method (i) lead to one solution while method (ii) yields two?

♣

Method (i) leads to the equation $0 = (5u - 2v)(5u + 2v)$ while (ii) leads to $4u^2 = 9u^2 + v^2 - 9uv/2$ or $0 = (5u - 2v)(2u - v)$. Let us examine more closely the second answer that is provided by (ii). In this case, the triangle formed by the two vectors and their resultant is isosceles with sides of magnitude $3u$, $2u$ and $2u$ with the base angle equal to θ and the apex angle equal to $180° - 2\theta$. In this configuration, $\cos\theta$ is indeed $3/4$, but the apex angle is not 2θ as specified in the statement of the problem. Indeed, method (i) made use of the angle between $\mathbf{R}$ and $\mathbf{G}$. However in method (ii) the result of the sine law remained valid with $180° - 2\theta$ in place of 2θ but the cosine law did not make use of the angle between $\mathbf{R}$ and $\mathbf{G}$. So it is not surprising that the second method led to a spurious possibility.

Solutions contributed by Don Curran of Oshawa, ON.

17. A linear pythagorean theorem

Figure 3.17 can be used to show that the sum of the lengths of the arms of a right triangle is equal to the length of the hypotenuse. (This with the more usual form of the pythagorean theorem leads to the result that $\sqrt{a^2 + b^2} = a + b$ for positive a and b.) We can approximate the hypotenuse ever more closely by a sequence of jagged curves, each of which has total length $a + b$.

This paradox appears in a number of places. Apparently, it was presented as a joke at the Collège de Beauvais to a class including the famous mathematician Lebesgue. While his classmates did not take it seriously, Lebesgue

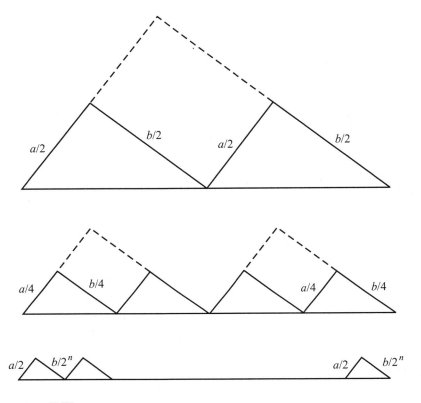

FIGURE 3.17

said "To me, the argument appears most disturbing, since I could see no difference between it and proofs relating to the areas and surfaces of cylinders, cones, spheres, and to the length of a circumference."

Contributed by George Mackiw and Christopher Morrell of Loyola College in Baltimore, MD. References supplied through the math-history-list@maa.org by John Conway, Roger Cooke, Mark McKinzie and Rick Otten.

References

1. C.B. Allendoerfer and C.O. Oakley, *Principles of mathematics*. McGraw Hill, New York, 1955 (11.2.2, p.288), 1963 (10.2.2, p. 324), 1969 (12.5.6, p. 465).
2. H.E. Dudeney, *Canterbury Puzzles*. Dover, Mineola, NY, 1958.

3. Kenneth Goldberg, Does the formula for arc length measure arc length? in Warren Page et al., eds. *Two Year College Mathematics Readings*, MAA, 1981: pp. 107–110.
4. Eugene P. Northrop, *Riddles in Mathematics: A Book of Paradoxes*. Van Nostrand, New York, 1944: p. 138.
5. Warren Page, The formula for arc length does measure arc length. in Warren Page et al., eds. *Two Year College Mathematics Readings*, MAA, 1981; pp. 111–114.
6. C. Tukey, *Nonstandard Methods in the Calculus of Variations*. Longman House, White Plains, NY, 1973.
7. Laurence C. Young, *Lectures on the Calculus of Variations and Optimal Control Theory*. W.B. Saunders, Orlando, FL, 1981: pp. 152–154.
8. Laurence C. Young, Mathematicians and their times. *Notas de Mathematica* 76 (1981) 142, 303.

18. The surface area of a sphere

To find the area of the northern hemisphere, divide the equator into n equal parts by points $A_1, A_2, \ldots, A_n$, and join these points to the North Pole N by arcs of meridians. Imagine now that the polygon $A_1 A_2 \cdots A_n$ starts to rise over the equatorial plane, staying parallel to it and contracting on its way so that its vertices slide along the meridians. Then its sides will cover the surface much like a closed bud. If the bud opens, we get n triangles. Let a_n be the base $(A_i A_{i+1})$ of any of the triangles and h_n the height. Then the total area of all the triangles equals $n a_n h_n / 2$. It is clear that as n increases, the area of the bud tends to the area of the hemisphere, while the polygon's perimeter $n a_n$ tends to the equator's length $2\pi R$ and the height to $\pi R/2$. Thus, the area of the hemisphere, which is the limit of the bud area, equals

$$2\pi R \cdot (\pi R/2) \cdot (1/2) = \pi^2 R^2 / 2.$$

Thus the surface area of a sphere equals $\pi^2 R^2$. ◊

This is from the article, "In search of a definition of surface area" by Vladimir Dubrovsky in *Quantum* 1:4 (1991) 6–9, 44.

19. Drenching a sphere

Here is a "Challenge Question" culled from page 132 of the text *Intermediate algebra with early functions*, 2nd edition (PWS Publishing Co., Boston, 1995) by James W. Hall.

A metal sphere is placed in a cylindrical beaker marked in cubic centimeters on the side. Approximate the radius of this sphere.

Accompanying this question is a pair of illustrations. In the first, the sphere rests beside a beaker filled to the 25 cc mark with water; in the second, the sphere is inside the beaker, enveloped by water extending to the 90 cc mark. ♣

Suppose that we had a sphere of radius r and volume $V = (4/3)\pi r^3$. What is the minimum amount of water to cover it in a cylindrical beaker? The radius of the cylinder is at least r and its height is at least $2r$, so its volume is at least $2\pi r^3$. Hence to cover the sphere, we need at least $(2 - 4/3)\pi r^3 = \frac{1}{2}V$ cubic units of water. In the given situation, the volume of the sphere is presumably 65 cc, so we would need at least 32.5 cc of water to cover it, which is more than the 25 cc in the first illustration.

Contributed by David Cantrell of Tuscaloosa, AL.

20. Volume of a tin can

In her regular *Parade Magazine* column (April 21, 1996), Marilyn Vos Savant fielded a question from a reader who wondered why we see so few rectangular cans and so many cylindrical ones. He surmised that rectangular cans take up less space and so would be more practical. Marilyn's response was that "a cylindical can requires less surface area of metal than a rectangular can to contain the same volume." This is not the whole story. Undoubtedly, Marilyn had in mind a comparison of cans of the same volume *and* height. In this case, the areas of the bases and lids of both cans would be equal and so only the vertical surface areas would need to be compared. With V and h, respectively, the volume and height of each can, a and b the base dimensions of the rectangular can, and r the radius of the cylindrical can, the difference between the surface areas is

$$2(a + b)h - 2\pi rh \geq 2h[2\sqrt{ab} - \pi r] = 2h\sqrt{\frac{V}{h}}[2 - \sqrt{\pi}] > 0,$$

so that, indeed, the cylindrical tin is more economical.

Suppose that we do not require equal heights. Let both cans have volume π and the dimensions of the rectangular can be $1 \times 1 \times \pi$. The surface area of the rectangular can is $2 + 4\pi$. If the cylindrical can has radius r, its height must be r^{-2} and its surface area $2\pi(r^2 + r^{-1})$, a quantity that can be made as large as desired.

Gleaned from Media Clips, a department edited by Ron Lancaster in the NCTM journal Mathematics Teacher 90 (1997) 119–120.

21. The Puptent Problem

The following question was number 44 in 3CPT1, a preliminary scholastic aptitude test set by the Educational Testing Service of New Jersey in 1980:

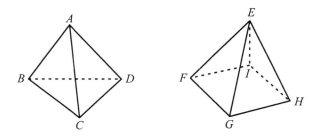

44. *In pyramids $ABCD$ and $EFGHI$ shown above, all faces except base $FGHI$ are equilateral triangles of equal size. If face ABC were placed on face EFG so that the vertices of the triangles coincide, how many exposed faces would the resulting solid have?**

(A) Five (B) Six (C) Seven (D) Eight (E) Nine

Apparently, the answer should be (C); the two solids together account for $5 + 4 = 9$ faces, but we have to subtract two since a face from each solid is concealed in the juxtaposition. This evidently was the opinion of the examiners, so that the candidates who selected (A) were originally credited with an incorrect answer.

In fact, there are only *five* faces on the combined solid. After the connection is made, two faces of the square pyramid turn out to be coplanar with corresponding faces of the tetrahedron.

There is a detailed discussion of this problem by Robert B. Davis on pages 219–235 of his book, *Learning mathematics: the cognitive science approach to mathematics education*, published in 1984 by Ablex of New Jersey. He established what he calls a "cognitive existence theorem," to wit that there is a representation for which the correct answer is immediately obvious. It is not productive to visualize both solids as sitting on the base figure. However, if one imagines two opposite edges of a tetrahedron lying in parallel planes but pointing in perpendicular directions, then the situation clarifies. For example, we might imagine that the pyramid $EFGHI$ is part of a "puptent" that has its slanted walls extended by ABD and ACD, anchored

---

* Reprinted by permission of Educational Testing Service and the College Entrance Examination Board, the copyright owners. For limited use by the Mathematical Association of America.

by attaching BC along FG and held up by a horizontal pole AD attached to E. An alternative visualization is to imagine a pair of square pyramids abutting along a base edge; the tetrahedron just fits in between the two, having one of its edges joining the two apexes to form a triangular prism. Davis refers to a 1982 paper by Young in Volume 3, Number 2 of the *Journal of Mathematical Behaviour* (pages 123–144).

I am grateful to Lawrence S. Braden, Robert B. Davis and John Kenelly on the internet for their assistance is helping me source this item.

22. The spirit is willing but the ham is rotten

John Kinloch and Rick Norwood presented the following "proof" of the Ham Sandwich Theorem in the *American Mathematical Monthly* 101 (1994) 470.

Theorem. *Given any sandwich composed of bread, ham and cheese, there is a single plane that cuts the sandwich into two parts, such that the two parts contain equal amounts of bread, equal amounts of ham, and equal amounts of cheese.*

Proof.   Let p, q, r, respectively, be the centers of mass of the bread, the ham, and the cheese. The plane that contains p, q and r divides the sandwich into two parts containing equal amounts of each of the bread, ham and cheese. ♠

Since the position of the centers of mass depends on the distribution of mass, it is not necessarily the case that a plane through the center of mass divides the mass into two equal parts. As a counterexample consider a triangular lamina and a plane perpendicular to the triangle that passes through its centroid and is parallel to a side.

Other items.   See also Item 2 (A superficial volume problem) in Chapter 2; Item 13 (Why Wiles' proof of the Fermat conjecture is wrong) in Chapter 1.

Chapter **4**

FINITE MATHEMATICS

Mathematical reasoning has to be pursued with great care, as there are pits that beset the unwary. We will begin with proofs by induction before going to other aspects of finite mathematics. Effecting a proof by induction is a sophisticated procedure that many students find quite mysterious. As long as they harbour the suspicion that somehow they are assuming what they have to prove, they are likely to treat it as a rote process and fall into confusion.

Perhaps the typographical error in the running head for page 589 of Michael Sullivan's *College Algebra* (Prentice Hall, 1995) says it all. Section 9.4, which treats mathematical induction, is headed *Mathematical Indirection.*

I. Rabbits reproduce; integers don't

Many readers will be familiar with some form of the following argument that all rabbits are the same color:

> Clearly one rabbit has the same color. Suppose any set of n rabbits has the same color, say white, and consider a set of $n + 1$ rabbits. Remove one rabbit from the set; the induction hypothesis tells us that the remaining n rabbits are white. To see that the removed rabbit is also white, put it back in the set and remove some other rabbit, obtaining another set of n rabbits, which by the induction hypothesis must also be white. Thus all rabbits are white. ◇

It is difficult to assess the mental demeanour of the upper division computer science student who provided this analysis:

> The faulty logic in the rabbit problem has to do with the behavior of rabbits and integers. For example, a set of rabbits is not isomorphic to a set of integers because their behavior is different. A set of integers is static while a set of rabbits is dynamic, i.e., integers

63

don't reproduce themselves while rabbits do. Since the two sets are not isomorphic, we cannot expect the same laws to apply to both sets. Since we know that the law of induction applies to a set of integers, we cannot expect the same law to apply to a set of rabbits that is not isomorphic. In the same manner, we cannot expect laws of reproduction which apply to a set of rabbits to apply to a set of integers.

Contributed by Annie and John Selden of Tennessee Technical University in Cookeville.

2. All positive integers are equal

To prove that all positive integers are equal, we first prove by induction that if the maximum of two positive integers is n, then the integers are equal. This is clearly true when $n = 1$. Suppose that it is true for $n = k$. Let u and v be two integers, the maximum of which is $k + 1$. Then the maximum of $u - 1$ and $v - 1$ is k, so by the induction hypothesis, $u - 1 = v - 1$. Hence $u = v$. With this result in hand, let two positive integers be given. Their maximum is some number n, so by the result, they are equal. $\Diamond$

This fallacy by T.I. Ramsamujh appeared in the *Mathematical Gazette* 72 (1988) 113 (item 72.14). Katalin Bencsath of Manhattan College, Riverdale, NY found an earlier appearance of the paradox in the 1984 first edition of the textbook *Discrete Mathematics* by Richard Johnsonbaugh (published by Macmillan) on page 38 as Exercise 27 in Section 1.5, entitled *Mathematical induction*. The exercise has been retained in both the "Revised edition" (1986) and "Second edition" (respectively #36 on page 59 and #39 on page 53).

3. Every second square is the same

Proposition. *For each positive integer n, $(n - 2)^2 = n^2$.*

Proof. Denote the given statement by S_n. Statement S_1 is true and forms the basis of the induction. Assuming S_k is true, let us examine S_{k+1}:

$$(k - 1)^2 = (k + 1)^2.$$

Expand to obtain $k^2 - 2k + 1 = k^2 + 2k + 1$, from which $-4k = 0$.

Since $k \geq 1$, we can multiply by $(1 - 1/k)$ to get $-4k + 4 = 0$. Add k^2 to both sides to get $k^2 - 4k + 4 = k^2$, from which we conclude $(k - 2)^2 = k^2$. Since S_{k+1} reduces to the known statement S_k, we may conclude that S_{k+1} is also true. Thus S_n has been shown to be true for all $n \geq 1$. $\heartsuit$

Contributed by Allen J. Schwenk of Western Michigan University in Kalamazoo.

4. Four weighings suffice

Problem.   Suppose you are given n (at least 2) coins that look identical, but one of which weighs less than the others, which are of equal weight. Show that the odd coin can be determined with no more than four weighings on an equal-arms balance.

Solution.   It is easy to see that a single weighing will suffice with two coins. Suppose that the result holds for $n = k$ coins, where $k \geq 2$. If we have $k + 1$ coins, we begin by laying one aside and applying the process to the remaining k coins. If we do not determine the odd coin among the remaining k coins, then the coin set aside must be it. ♠

Observe that the number "four" plays no role in the solution; indeed with two coins one can get by with a single weighing, and so the argument should wash even if "four" is replaced by "one." But let us examine the induction step more closely. Having laid aside one of the $k + 1$ coins, we have to deal with the remaining k coins. But, according to the process, we now lay aside one of these k coins. If no distinction is found among the $k - 1$ coins still in play we now have two set-aside coins to adjudicate between and this will require another weighing.

Argument of Keith Austin from the Mathematical Gazette 72 (1988) 113, item 72.15.

5. Perron's paradox

Proposition.   *1 is the largest positive integer.*

Proof.   Let n be the largest positive integer. Since $n^2 \geq n$, we must have $n^2 = n$. But then $n = 1$. ♡

Reference to this paradox is made in *Dirichlet's Principle: A Mathematical Comedy of Errors and its Influence on the Development of Analysis* by A.F. Monna (Oosthoek, Scheltema & Holkema, Utrecht, 1975) in the context of making the assumption of an optimizing curve in the proof of Steiner's theorem that among all planar closed curves of a given length, it is the circle that encloses the largest area; one argues that, given any noncircle, we can always find a better curve. This issue also arises in problems of the calculus of variations, in particular in the application of Dirichlet's principle.

Monna draws attention to the reference *Les problèmes de isopérimètres et des isépiphanes*, by T. Bonnesen (Paris, 1929). A letter appearing in the journal *Eureka (Crux Mathematicorum)* 3 (1977) 187–188 compares this argument with a similar one to establish that among all triangles inscribed in a given circle, the equilateral has the largest area.

6. There is a unique positive integer

This is not an induction argument, but it does have a similar flavor to what we have been considering. Many students would probably accept that the proof is faulty, but it is such a muddle that few would have the mathematical experience or analytic skill to take it apart.

Proposition. *If* $1 + 2 + \cdots + n = n(n+1)/2$ *for all positive integers* n, *then* $n = 1$.

Proof. Replacing n with $n - 1$ in the hypothesis gives

$$1 + 2 + 3 + \cdots + (n-1) = \frac{n(n-1)}{2} \qquad \text{for } n \geq 2.$$

Add 1 to both sides of the latter identity to yield

$$1 + 2 + 3 + \cdots + n = \frac{n(n-1)}{2} + 1.$$

Comparing this with the hypothesis yields $\frac{1}{2}n(n+1) = \frac{1}{2}(n-1)n+1$ which simplifies to $n = 1$. ♡

7. A criterion for a cyclic graph

With most faulty induction proofs, the glibly concealed fallacy or oversight in the induction step occurs right away, or possibly at the second or third step. Here is an argument that starts at $n = 1$ and first fails at $n = 6$. The reader is reminded of a few definitions. A *graph* is a collection of points, called *vertices*, and arcs or curves, called *edges*; each edge has two ends, and each of its ends is a vertex. An edge whose two ends are the same is called a *loop*. The *degree* of a vertex is the number of times it occurs at the end of an edge. A *simple graph* is a graph with no loops such that, for any two different vertices, there is at most one edge with those vertices as its ends. A *cycle of length one* is a loop; a *cycle of length two* consists of a "doubled edge," i.e., a graph consisting of two edges, two vertices and no loops. For

$n > 2$, a *cycle of length* n is a graph that looks like a polygon with n sides and n vertices.

Theorem. *If G is a finite simple graph in which every vertex has degree two, then G is a cycle.*

Proof. The proof is by induction on the number n of vertices of G. In a simple graph on n vertices, there is no vertex of degree exceeding $n - 1$; therefore there are no graphs G satisfying the hypothesis when $n = 1$ or $n = 2$; when $n = 3$ the only one is the complete graph K_3 consisting of three vertices, each pair joined by an edge, and this is a cycle.

Assume that $n > 3$. Let u be a vertex of G. In a simple graph the degree of a vertex is equal to the number of its neighbors, so u has exactly two distinct neighbors, say v and w. Now let $H = (G - u) + vw$; this notation means that H is obtained by deleting u and the two edges incident with it, and then adding an edge between v and w. Thus, H has $n - 1$ vertices, all of degree two. By the induction hypothesis, H is a cycle. G can be obtained from H by inserting a vertex (let us call it u!) on the edge between v and w referred to above. Clearly, performing this operation on a cycle results in a cycle, so G is a cycle. ♠

The graph which is the union of two disjoint copies of K_3 (a triangle) is a counterexample. For any vertex u with two neighbors v and w, the vertices of one copy of K_3 are u, v and w. Deleting u and its adjacent edges, and inserting an edge between v and w would give two separate edges connecting v and w, and so yield a nonsimple graph beyond the scope of the induction hypothesis.

Faulty argument constructed by P. D. Johnson and Martin Schlam of Auburn University in Alabama.

8. Doggedly bisexual

On page 27, the British booklet *Investigating Sets* by Ed Catherall (Wayland, Hove, UK, 1982) offers its young readers a different perspective on the intersection of two sets:

List the names of your friends that are children. Put your friends that are boys in a boys' set B. Put your friends that are girls in a girls' set G.

Which of your friends have dogs as pets? Notice that both boys and girls have dogs as pets. Show this as a Venn diagram with the

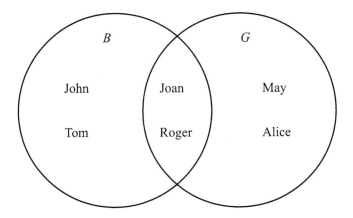

FIGURE **4.8**

boy and girl sets connecting. In the overlap write the names of the boys and girls that have dogs as pets.

The symbol for the connection or intersection is ∩.

$$B = \{\text{John, Tom, Roger}\}$$

$$G = \{\text{May, Alice, Joan}\}$$

If Joan and Roger have dogs then

$$B \cap G = \{\text{Joan, Roger}\}. \lozenge$$

Contributed by Neal Madras of York University in Toronto, ON.

The next two deliberately erroneous items are from *Foundations of Higher Mathematics: Exploration and Proof* by Daniel Fendel and Diane Resek (Addison-Wesley, Reading, MA, 1990).

9. Equal unions

Theorem.   *For any sets A, B and C, if $A \cup B = A \cup C$, then $B = C$.*

Proof.   Assume that $A \cup B = A \cup C$. We need to prove $B = C$. First we will prove that $B \subseteq C$. Suppose that x is in B. Then x is certainly in $A \cup B$. By assumption, therefore, x is also in $A \cup C$. But we did not assume x was in A, so x must be in C. Thus, we have shown that $B \subseteq C$. By a parallel argument, $C \subseteq B$. ♡

10. Surjective functions

Theorem.   *For $f : A \to B$ and $g : B \to C$, if $g \circ f$ is surjective, then f is surjective (i.e., onto).*

Proof.   Assume that $g \circ f$ is surjective. Let $t \in B$. We have $g(t) \in C$. Since $g \circ f$ is surjective, there is an element r in A such that $(g \circ f)(r) = g(t)$. But if $g(f(r)) = g(t)$, then $f(r) = t$. Therefore we have found the desired element r. $\heartsuit$

11. Hockey ranking

The East Coast Hockey League has an unusual way of ranking teams. If a team wins a game in the regulation 60 minutes of play, it receives two ranking points while the loser receives no points. If a game is tied at the end of regulation, then a winning team is determined by a "shootout" or sequence of penalty shots for each team. The winning team again receives two points. However, now there is no losing team; the nonwinning team receives one point and is credited with a tie.

One result of this point allotment method is that interpreting win-loss-tie records can be tricky. For example, in the 1996–97 season South Carolina had a record of 3-4-3 versus Richmond (records are always given as wins-losses-ties). One might think that South Carolina had done almost as well as Richmond in the series of games between the two teams; but Richmond's record against South Carolina was 7-3-0. Richmond with 14 points had done much better than South Carolina with 9. You may notice another quirk of the method: from South Carolina's record against Richmond, one cannot determine Richmond's record against South Carolina; all we know is that Richmond had seven wins.

Teams can benefit from shootout games. For example, suppose teams A, B and C all play nine games against each other. Suppose that the results are as follows:

Competitors	Record of A	Record of B	Record of C
(A, B)	4-4-1	5-4-0	——
(B, C)	——	5-4-0	4-4-1
(A, C)	5-2-2	——	4-1-4

Thus, in the match between A and B, there was one shootout game won by B, while in the match between A and C, C won two of the three games that ended in regulation and A won four out of six shootout games. The combined records for the three teams are

$$A : 9\text{-}6\text{-}3 \qquad B : 10\text{-}8\text{-}0 \qquad C : 8\text{-}5\text{-}5.$$

Thus A and C tie for first with 21 points, while B is last with 20 points despite having outperformed each team in head-to-head play. In practice, this type of thing might occur if the league would allot many of its 16 playoff positions to wildcard teams. Then a weak division might have more than its fair share of wildcard teams if for some reason a high percentage of its intradivisional games went into shootout. (A wildcard team is one that makes the playoffs by having the best record among the teams not heading their divisions.)

Contributed by Dave Trautman of The Citadel in Charleston, SC.

12. Spoiled for choice

This item appeared in the Sports Section (page 14) of the Sunday, February 22, 1998 edition of *The New York Times*:

> Big pickups also appeal because of the seemingly infinite ways they can be personalized; you need the math skills of Will Hunting to total the configurations. For starters, there are 32 combinations of cabs (standard, Club Cab, Quad Cab), cargo beds (6.5 or 8 feet) and engines (3.9-liter V6, 5.2-liter V8, 5.9-liter V8, 5.9-liter turbodiesel inline 6, 8-liter V10). You can have the basic ST, the fancy Laramie SLT or the drab-as-Monday-morning Work Special.

Contributed by Norton Starr of Amherst College in Massachusetts, who comments that it is cheap to blame the gaffe on the fictional Will Hunting; the author here totaled this computation all on his own.

13. Arranging a collection

The argument that the number of arrangements of a class of n students in a room with exactly n chairs can be effected in $(n!)^2$ ways follows:

> We first select a student, which can be done in n ways. Then we must select a chair for this student which can also be done in n ways. Thus, the first student is selected and seated in n^2 ways. Next, we select the second student in $n-1$ ways and his place in $n-1$ ways, so that the selection and seating of the second student can be done in $(n-1)^2$ ways. Continuing on in this way, we select and seat all the students in $n^2(n-1)^2(n-2)^2 \cdots 3^2 \cdot 2^2 \cdots 1^2 = (n!)^2$ ways.♡

The answer would be correct for the following problem: the students are selected one after another and the chairs assigned, where we distinguish among different orders of selection (even if the assignment of chairs is the same for all the students).

Contributed by Montie Monzingo of Southern Methodist University in Dallas, TX.

I4. A full house

Here are two separate solutions of drawing from a standard deck of 52 cards a "full house" consisting of 5 cards with three of one rank and two of another.

Solution 1. There are $13 \cdot 12$ ways to pick 2 different ranks, one to be "three of a kind" and the other to be a "pair." For each of these, there are $4!/3!1!$ ways to get the "three of a kind" and $4!/2!2!$ ways to get the "pair," so the probability of drawing a full house is

$$\frac{13 \cdot 12 \cdot \dfrac{4!}{3!1!} \cdot \dfrac{4!}{2!2!}}{\dfrac{52!}{5!47!}} = \frac{3744}{2598960} = 0.00144.\heartsuit$$

Solution 2. Consider first how many four-card hands there are that consist of two different pairs. There are $13!/2!11!$ ways to designate the two different denominations and for each of these there are $4!/2!2!$ ways to get one of the pairs and $4!/2!2!$ ways to get the other pair. So there are

$$\frac{13!}{2!11!} \cdot \frac{4!}{2!2!} \cdot \frac{4!}{2!2!}$$

different four-card hands consisting of two pairs. Now for each of these hands there are four cards that turn those two pairs into a full house, so the probability of drawing a full house is

$$\frac{\dfrac{13!}{2!11!} \cdot \dfrac{4!}{2!2!} \cdot \dfrac{4!}{2!2!} \cdot 4}{\dfrac{52!}{5!47!}} = \frac{11232}{2598960} = 0.00432.\spadesuit$$

The first solution is correct. In the second solution, each favorable hand gets counted three times. For example, the hand

$$J\spadesuit \quad J\heartsuit \quad J\diamondsuit \quad Q\spadesuit \quad Q\heartsuit$$

appears as a case for each of the following choices of pairs:

$$J\spadesuit \quad J\heartsuit \quad Q\spadesuit \quad Q\heartsuit$$
$$J\heartsuit \quad J\diamondsuit \quad Q\spadesuit \quad Q\heartsuit$$
$$J\spadesuit \quad J\diamondsuit \quad Q\spadesuit \quad Q\heartsuit$$

Contributed by Eric Chandler of Randolph-Macon Woman's College in Lynchburg, VA.

15. Which balls are actually there?

In a well-conceived article on how children and adults comprehend the concept of infinity [*Theory & Psychology* 4:1 (1994) 35–60], Ruma Falk poses and analyzes the following conundrum:

> An infinite line of tennis balls, numbered $1, 2, 3, \ldots$ is arranged in front of an empty room. Half a minute before 12 o'clock, balls #1 and #2 are thrown into the room and #1 is thrown out. A quarter of a minute before 12:00, balls #3 and #4 are tossed in and #2 is tossed out. In the next $\frac{1}{8}$- minute, #5 and #6 are tossed in and #3 is tossed out, and so on. The question is, how many balls will the room contain at 12:00? Cogent arguments can be made for two diametrically different answers.

> *First answer.* There will be infinitely many balls in the room at 12:00. Argument 1: The number of balls in the room increases by one at each tossing event. Hence for any N you suggest, I can compute an exact time (before 12:00) when the number of balls in the room has exceeded N.

> *Second answer.* There will be no balls in the room at 12:00. Argument 2: If you claim that there is any ball there, when you name it, I can tell you the exact time (before 12:00) when it was tossed out.

A probabilistic version of this situation is described in the text. *A first course in probability,* by Sheldon Ross (5th edition, Prentice-Hall, 1998). In Example 6a of Section 2.6, he describes putting in ten balls at a time and removing one. When the one is removed at random, then the probability that a given ball is in the bag at 12 o'clock is 0.

16. Red and blue hats

A group of men, all versed in logic, sits around a table. Fifteen of them are wearing red hats, the rest blue. Each can see the others' hats but not his own. On the table is a clock which strikes once each hour. The men are given the following instructions: "You are not allowed to discuss the color of your hats. However, should any of you find that he is wearing a red hat, he should leave the table on the clock-strike immediately following his discovery." Now it is assumed that no one is initially aware of the color of his own hat. Furthermore, since the men cannot see the color of their own hats, nor discuss it with their colleagues, nothing happens for a while. Then a guest arrives. She looks at the hats around the table, and says clearly: "At least one man here is wearing a red hat!" What happened and why?

An induction argument can be used to show that once the guest has made her announcement, all individuals with red hats will eventually leave the table. This contradicts one's feeling that nothing should continue to happen; after all, apparently the guest has not provided any new information.

A version of this puzzle appears in Michael Spivak, *Calculus* (second edition) Publish or Perish, Inc., page 35, problems 27 and 28. What the guest contributes to the situation is touched on in Spivak and also discussed by Uri Leron and Mike Eisenberg in their paper, "On a knowledge-related paradox and its resolution," *Int. J. Math. Educ. Sci. Technol.* 18 (1987) 761–765. In a subtle fashion, Leron and Eisenberg distinguish between facts known to each individual and facts that become public knowledge within the group.

I am indebted to a reviewer for directing my attention to the charming children's book, *Anno's hat tricks*, by Akihiro Nozaki and Mitsumasa Anno (Philomel Books, 1985) ISBN 0-399-21212-4. This is an adventurous but gentle introduction for the young to logical reasoning, in which the author puts the reader into the problem as a "Shadowchild" whose shadow flits across the page and invites her to engage in reasoning with two other children, Tom and Hanna, as they determine the colors of their hats, in situations ranging from the simplest up to having three red and two white hats, three of which are placed on the heads of the children. In an essay entitled "Mathematical induction of colored hats," Martin Gardner also discusses the same problem. This is chapter 10 in his book, *Penrose tiles to trapdoor ciphers ... and the return of Dr. Matrix* (MAA, Washington, DC, 1997).

A similar problem was posed as #3734 in the *American Mathematical Monthly* and published as one of the four hundred "best" problems for the period 1918–1950 in *The Otto Dunkel Memorial Problem Book* (MAA, 1957):

A car with n $(n > 2)$ passengers of different speeds of mental reaction passes through a tunnel and each passenger acquires unconsciously a smudge of soot upon his forehead. Suppose that each passenger

 (1) laughs and continues to laugh as soon as and only as long as he sees a smudge upon the forehead of a fellow passenger;
 (2) can see the forehead of all his fellows;
 (3) reasons correctly;
 (4) will clean his own forehead when and only when his reasoning forces him to conclude that he has a smudge;
 (5) knows that (1), (2), (3), (4) hold for each of his fellows.

Show that each passenger will eventually wipe his own forehead.

17. An invalid argument

In what follows, p and q refer to propositions, $\vee$ and $\wedge$ are the respective connectives *or* and *and,* and $\neg$ means *not.* The triangle of dots means *therefore.*

Proposition.   The argument

$$q \vee \neg p$$
$$\underline{\qquad \neg q \qquad}$$
$$\therefore p$$

is invalid.

Proof.   Suppose the argument were valid. It can also be shown that the argument

$$q \vee \neg p$$
$$\underline{\qquad \neg q \qquad}$$
$$\therefore \neg p$$

is valid. To see this, one first notes that $q \vee \neg p$ is equivalent to $p \rightarrow q$, and then uses the contrapositive together with $\neg q$ to conclude $\neg p$. Thus one arrives at $p \wedge \neg p$, which is a contradiction. Hence, the supposition is untenable and the original argument is shown to be invalid. $\heartsuit$

The above line of reasoning does not establish the original argument as invalid—one has not "arrived" at $p \wedge \neg p$. But the approach does suggest something. If

$$P$$
$$\frac{Q}{\therefore R} \quad \text{and} \quad \frac{Q}{\therefore \neg R}$$

are both valid, then

$$P$$
$$\frac{Q}{\therefore R \wedge \neg R}$$

is also valid. Hence $P \wedge Q$ is logically false and

$$P$$
$$\frac{Q}{\therefore S}$$

is valid for any S.

Contributed by Annie Selden of the Tennessee Technological University and John Selden of MERC, both in Cookeville, TN.

18. A logical paradox

Proposition. Either an implication or its converse must be true.

First proof. Consider the truth table based on the form $(p \rightarrow q) \vee (q \rightarrow p)$:

p	q	$p \rightarrow q$	$q \rightarrow p$	$(p \rightarrow q) \vee (q \rightarrow p)$
T	T	T	T	T
T	F	F	T	T
F	T	T	F	T
F	F	T	T	T

Hence $(p \rightarrow q) \vee (q \rightarrow p)$ proves a tautology and is accordingly true regardless of the truth values assigned to the variable parts p and q. ♡

Second proof. $p \rightarrow q$ is equivalent to $q \vee \neg p$. Thus

$$(p \rightarrow q) \vee (q \rightarrow p) = (q \vee \neg p) \vee (p \vee \neg q) = q \vee \neg p \vee p \vee \neg q = 1. \spadesuit$$

Richard L. Francis of Southeast Missouri State University in Cape Giraudeau, who submitted these arguments, points to a consequence: Taking p and q appropriately, one has that "if a number is prime, then it is odd, or if a number is odd, then it is prime."

Carl Jongsma of Dordt College, Sioux Center, Iowa comments:
The paradox hinges on at least two confusions. The first one blurs
logical syntax and semantics. Although $(p \rightarrow q) \vee (q \rightarrow p)$ is
a tautology under the conventional truth value definition for $\rightarrow$,
which means for any sentences p and q that either $p \rightarrow q$ or $q \rightarrow p$
is true, we may not conclude that either "p implies q" is true or "q
implies p" is true. Logical implication cannot be captured by this
or any other truth functional connective. Thinking that it can leads
to several paradoxes of implication such as this one.

Secondly, a universal quantifier has been illegally distributed
in the particular example provided to make the given statement seem
more paradoxical. Since the sentence "if n is prime, then n is odd;
or if n is odd, then n is prime" is always true, the universal closure,
"for all natural numbers n, if n is prime, then n is odd; or if n is
odd, then n is prime" is also true. However, the distributed universal
disjunction "for all natural numbers n, if n is prime, then n is odd;
or for all natural numbers n, if n is odd, then n is prime" need
not be true (and of course isn't). All numbers are even or odd, for
instance, but it is not the case that all of them are even or all of
them are odd. Thus it is invalid to distribute the universal quantifier
over a disjunction. Such errors result from failing to be clear about
the position of quantifiers in informal mathematical statements.

He points to the problem of "blurring the distinction between the seman-
tic notion of logical implication (not a truth functional operator on propositions, but a logical relation between them) and the conventional truth function
syntactic operator '$\rightarrow$'." Further, the conditional connective ought not to be
read as "implies." "The Deduction Theorem of propositional logic allows you
to get away with translating $\rightarrow$ as 'implies' (or 'proves') in certain contexts,
but not here. The confusion between '$\rightarrow$' and 'implies' is somewhat analogous
to confusing the operation of division with the binary relation of 'divides',
something students tend to do when they first meet the notation $a|b$."

Other items.   See also Chapter 2, item 18, and Chapter 6, item 1.

Chapter **5**

PROBABILITY

I. Meeting in a knockout tournament

Omicron and Upsilon are discussing Problem 297 from *Five Hundred Mathematical Challenges* (MAA, 1996):

> A tennis club invites 32 players of equal ability to compete in an elimination tournament. (This proceeds in a number of rounds in which players compete in pairs; any losing player retires from the tournament.) What is the probability that two given players will compete against each other?

Omicron. The solution seems straightforward enough. There have to be 31 games to knock out all but the ultimate winner. There are $\binom{32}{2}$ possible pairs, so that the probability of a given pair being selected for a particular match is $1/\binom{32}{2} = 1/16 \cdot 31$. Since the selection for the players in the different matches is mutually exclusive, the probability of a given pair being selected is 31 times this, which is $1/16$. ♣

Upsilon. Hold on! This presupposes symmetry among all the matches. The tournament proceeds in several rounds. In the initial round of 16 games, the pairs may indeed be selected at random. But in subsequent rounds, a pair is selected only if both players survive the previous round. Your argument does not use the fact that the players are of equal ability. Two players who are sure to beat everyone else in the tournament (say, Steffi Graf and Pete Sampras) will meet with probability 1; two players inferior to everyone else will meet only in the first round or not at all.

Omicron. But the fact that the players are of equal ability provides the symmetry that you crave. How would you do the problem?

Upsilon. This seems to be a situation for an induction argument. After the first round, we have the same situation for 16 players, after the second for

8 players, and so on. Let p_k be the probability that a given pair will meet if we start with 2^k players of equal ability. Clearly $p_1 = 1$. If we start with 4 players, a given pair will meet in the first round with probability 1/3 and vanquish other players to meet in the second and final round with probability $(2/3)(1/2)(1/2) = 1/6$. Thus $p_2 = 1/2$. In general, we find that

$$ p_k = \frac{1}{2^k - 1} + \left[1 - \frac{1}{2^k - 1} \right] \left(\frac{1}{2} \right)^2 p_{k-1}, $$

the first term being the probability of the pair meeting in the first round and the second term of each person beating others to get to the second round and eventually meeting. Solving this recursion leads to... let me see... $p_k = 1/2^{k-1}$. The answer to the problem is 1/16.

Omicron.   I am vindicated!

Idea due to Ruma Falk of the Hebrew University of Jerusalem, Israel.

2. Where the grass is greener

There are two cards on the table. One them has written on it a positive number; the other, half that number. One of the cards, selected by a coin flip, is revealed to you. You may get in dollars either the number on this card or the number on the other card. Which should you choose?

Suppose that the number revealed to you is A. Then the other card has the number $2A$ or $0.5A$, each with equal probability. If you stick with the card shown, your expected winnings are A. If you switch, then your expected winnings are $0.5(2A) + 0.5(0.5A) = 1.25A$. Thus, you should always select the card other than the one that was revealed to you. ♣

This drew responses from three readers of the *College Mathematics Journal*. Samuel Goldberg of New York, NY, said that the reasoning was faulty. Based on a comparison of expected winnings, it implicitly assumes that the decision-maker is risk neutral. A more realistic analysis would introduce the decision-maker's utility for money. Risk-averse or risk-prone behavior can thus be modelled and the decision that "should" be taken customized to the decision-maker by maximizing expected utility rather than expected monetary winnings. Those who wish to pursue these issues can consult a beginning book on decision analysis. Some possibilities are: Karl H. Borch, *The economics of uncertainty*, Princeton, 1968; R.D. Luce and H. Raiffa, *Games and decisions*, Wiley, 1957; Anatol Rapoport, *Decision theory and decision behaviour: normative and descriptive approaches*, Kluwer, 1989.

A modified "double or nothing" version illustrates the difficulty. Suppose the choice is between the sure amount of A dollars or a lottery in which you get $2A$ dollars or nothing with equal probability. With either option, the expectation is A. For small A, many would try for "double or nothing." But for $A = 10^6$, most of us would take the money rather than risk ending up with nothing. Comparing expected monetary winnings is not satisfactory in deciding how one "would" or "should" act in choices under uncertainty.

On the other hand, David M. Bloom from Brooklyn College in New York takes an analytical approach. He suggests that one might wish to take the number shown. If the revealed number is large, then it is probably larger and should be selected; if it is small, then so is the difference and thus it does not matter what is chosen.

More mathematically, suppose to simplify the discussion, that the larger of the two numbers is a random variable having a continuous density function $g(x)$ on $(0, \infty)$. If x is the number on the revealed card, then standard arguments show that the probability $p(x)$ of x being the larger number is

$$p(x) = \frac{g(x)}{g(x) + 2g(2x)}.$$

Hence $p(x) = \frac{1}{2}$ for all x only if $g(x) = 2g(2x)$ for $x > 0$. This equation is satisfied by *no* continuous density function on $(0, \infty)$. Thus the assumption $p = \frac{1}{2}$ is false.

Another analysis is provided by S. L. Paveri-Fontana of Milan, Italy. Let the random variable X represent the number revealed and Y the larger of the two numbers. In general, for $c > 0$, we have $\Pr(X = c | Y = c) = \Pr(X = c | Y = 2c) = \frac{1}{2}$. By Bayes' Theorem,

$\Pr(Y = 2A | X = A)$

$$= \frac{\Pr(X = A | Y = 2A)\Pr(Y = 2A)}{\Pr(X = A | Y = 2A)\Pr(Y = 2A) + \Pr(X = A | Y = A)\Pr(Y = A)}$$

$$= \frac{\Pr(Y = 2A)}{\Pr(Y = 2A) + \Pr(Y = A)}.$$

Similarly,

$$\Pr(Y = A | X = A) = \frac{\Pr(Y = A)}{\Pr(Y = 2A) + \Pr(Y = A)}.$$

We have obtained the probabilities for the second card to carry $2A$ or $\frac{1}{2}A$, respectively, given that A is revealed. There is absolutely no reason to presume that such probabilities are equal.

If I do not switch, I collect A. Otherwise, my expected gain is

$$2A\Pr(Y = 2A | X = A) + \frac{1}{2}A\Pr(Y = A | X = A).$$

The difference in these gains is $A\Pr(Y = 2A|X = A) - \frac{1}{2}A\Pr(Y = A|X = A)$. I should switch iff this is positive. To decide, I need *a priori* knowledge on the distribution of $\Pr(Y = c)$ as a function of c. The naive instinctive strategy of switching when A appears small is not that bad after all. The trick of course is to decide what is "small."

As the last comment suggests, the difficulty arises because of a shift in perspective. Let K and $2K$ be the numbers written on the cards. The outcomes $A = K$ and $A = 2K$ are equally probable for the revealed card (as well as for the hidden card). Whether one adopts the strategy of always picking the revealed card or always picking the hidden card, the expectation is $1.5K$. This would turn out to be $1.5A$ or $0.75A$ depending on whether K or $2K$ is revealed. The original analysis equates the situation to one in which the amount A is seen and then one writes on the hidden card one of the amounts $\frac{1}{2}A$ or $2A$ as determined by a coin flip. If this happens, the expectation for the hidden amount is indeed $1.25A$.

The problem has had recent attention in the literature. It made an appearance in the *Parade* column "Ask Marilyn" of September 20, 1992, and was the subject of a "Reader Reflection" by Deborah Hecht in the *Mathematics Teacher* 85 (1992) 90–91. For other treatments, see

Steven J. Brams and D. Marc Kilgour, The box problem: to switch or not to switch. *Mathematics Magazine* 68 (1995) 27–34 (reviewed in *CHANCE News* 4.04 (16 February to 3 March, 1995)).

Ronald Christensen and Jessica Utts, Bayesian resolution of the "Exchange Paradox" *American Statistician* 46 (1992) 274–276.

Ruma Falk and Clifford Konold, The psychology of learning probability. In F. S. Gordon and S. P. Gordon, *eds.*, *Statistics for the twenty-first century,* MAA, Washington, DC, 1992; 151–164.

Elliot Linzer, The two envelope paradox. *American Mathematical Monthly* 101 (1994) 417–420.

Item contributed by Richard K. Guy of the University of Calgary in Alberta.

3. How to make a million

Consider this game. You and a friend take out your wallets and count your money. The person with the smaller amount of money will get the contents of both wallets. You reason as follows: I do not know how much money my friend has, but there should be a 0.5 chance that it is more than I have. So the probability that I will win this game is $\frac{1}{2}$. But if I win, I win the greater amount. If I lose, I lose the lesser amount. Therefore I should play this game as I expect to win more than I expect to lose. Of course your friend is reasoning the same way. ♣

For the sake of argument, let us suppose that no one is likely to possess more than k cents, but that each of the amounts $0, 1, 2, \ldots, k$ cents in the friend's pocket is equally likely. Then, if I have n cents $(0 \leq n \leq k)$, my expected winnings is

$$\frac{n}{k+1}(-n) + \frac{1}{k+1}(0) + \sum_{i=n+1}^{k} \frac{i}{k+1} = \frac{k}{2} - \frac{n(3n+1)}{2(k+1)}.$$

(This assumes that no money changes hands if the amounts are equal.) If the amount I have is randomized with each amount from 0 to k inclusive equally likely, it can be shown that my expectation is 0.

The assumption assigning probability $\frac{1}{2}$ to one person having more than the other is facile and coarse. More realistically, one should impute a probability distribution function which makes moderate sums more likely than either small or large amounts. Suppose $q(y)$ $(y > 0)$ is the probability density function for the amount of money in my friend's wallet. If I actually have x, then my expected gain is

$$f(x) = \int_x^\infty yq(y)\,dy - x\int_0^x q(y)\,dy \qquad (x > 0)$$

which is monotonically nonincreasing with $f(0) > 0$ and $f(\infty) = -\infty$. Clearly one expects to gain when x is small and to lose when x is large.

The problem is discussed by Laurence McGilvery in his article, "Speaking of paradoxes $\cdots$ or are we?" in the *Journal of Recreational Mathematics* 19 (1987) 15–19. He treats the situation in which each person is given a wallet and is ignorant of the sum it contains. A brief treatment of related paradoxes can be found on pages 145–148 of Martin Gardner, *Penrose tiles to trapdoor ciphers... and the return of Dr. Matrix* (MAA, Washington, DC, 1997) ISBN 0-88385-521-6.

Contributed by Larry Clevensen of California State University in Northridge. Analysis with density function due to Stefano Paveri-Fontana of Milano, Italy.

4. A problem of Lewis Carroll

Do you have trouble sleeping at night? Lewis Carroll apparently did on occasion, and spent the solitary hours creating and solving "pillow problems." These were eventually published, recently by Dover Publications. The famous Carrollian sense of anomaly is apparent in Problem No. 72:

Problem.   A bag contains 2 counters, as to which nothing is known except that each is either black or white. Ascertain their colors without taking them out of the bag.

Answer.   One is black, and the other white.

Solution.   We know that, if a bag contained 3 counters, 2 being black and one white, the chance of drawing a black one would be 2/3; and that any *other* state of things would *not* give this chance. Now the chances, that the given bag contains BB, BW, WW, are, respectively, 1/4, 1/2, 1/4. Add a black counter. Then the chances that it contains BBB, BWB, WWB, are, as before, 1/4, 1/2, 1/4. Hence the chance, of now drawing a black one, $= (1/4) \cdot 1 + (1/2) \cdot (2/3) + (1/4) \cdot (1/3) = 2/3$. Hence the bag contains BBW (since any *other* state of things would *not* give this chance). Hence, before the black counter was added, it contained BW, i.e., one black counter and one white. Q.E.F. $\heartsuit$

One can apply a similar argument to establish that if a bag contains one counter, either white or black, then it can only be empty. Add a black counter to the bag and select one at random. The probability that the selected one is black is $(1/2) \cdot 1 + (1/2) \cdot (1/2) = 3/4$. This does not correspond to the probability of drawing a black counter from a bag containing two whether both are black, both are white or both differ in color. The apparent anomaly arises because we are conflating two problems, one in which the random event is drawing a counter from a bag of known contents and one where the contents of the bag are partially randomized before the drawing is made.

5. Nontransitive dice

Let A, B, and C be three cubical dice whose faces are marked as follows:

$$A : 18, 9, 8, 7, 6, 5$$

$$B : 17, 16, 15, 4, 3, 2$$

$$C : 14, 13, 12, 11, 10, 1$$

Suppose that the dice are rolled. Each of the following probabilities exceeds $1/2$: the probability that the number on A exceeds that on B (21/36); the probability that B beats C (21/36); the probability that C beats A (25/36).

In his article, "The paradox of nontransitive dice" (*American Mathematical Monthly* 101 (1994) 429–436). Richard P. Savage, Jr., studies this paradox for the case of n-sided dice with the numbers from 1 to $3n$ inclusive and gets information about the minimum of the three probabilities.

6. Three coins in the fountain

Sir Francis Galton published the following comments in the February 15, 1894 issue of *Nature* (Volume 49, page 365).

It seems worthwhile to record the following pretty statistical paradox as a good example of the pitfalls into which persons are apt to fall, who attempt short cuts in the solution of problems of chance instead of adhering to the true and narrow road. It is true that the paradox would excite immediate suspicion in the mind of anyone accustomed to such problems, but I doubt there are many who, without recourse to paper and pen, could *distinctly* specify off-hand where the fallacy lies. It will be easy for the reader to make the experiment of his own competence to do so after reading to the end of the second of the two following paragraphs.

The question concerns the chance of three coins turning up alike, that is, all heads or else all tails. The straightforward solution is simple enough; namely, that there are 2 different and equally probable ways in which a single coin may turn up; there are 4 in which two coins may turn up, and 8 ways in which three coins may do so. Of these 8 ways, one is all-heads and another all-tails, therefore the chance of being all-alike is 2 to 8, or 1 to 4.

Against this conclusion I lately heard it urged, in perfect good faith, that as at least two of the coins must turn up alike, and it is an even chance whether a third coin is heads or tails; therefore the chance of being all-alike is as 1 to 2, and not as 1 to 4. Where does the fallacy lie? ♣

Suppose we distinguish the coins A, B and C. If A and B are alike, then there is an even chance of coin C matching them. A similar comment applies when A and C and when B and C are alike. The difficulty arises because the third coin being the same when A and B are alike coincides with the event of its being the same when A and C or when B and C are alike. This is not the case for the third coin being different.

Galton's erroneous argument is discussed by Eugene Northrop in his *Riddles in mathematics: a book of paradoxes* (pages 172–175). It is also analyzed in the article by Ruma Falk, A classic probability puzzle, *Teaching Statistics* 18:1 (1996) 17–19.

7. Getting black balls

The following problem was originally posed in 1980 in a Belgian mathematics competition. It was reproduced in the Toronto *Star* in July, 1995, as one of a set of problems posed to the readership at the time of the International Mathematical Olympiad. The solution and comment were provided by a reader.

Problem.   Each of two urns contains black and white balls, where the total number of balls in both urns is 25. One ball is drawn from each bag at random. The probability that both are white is 0.54. Find the probability that both are black.

Solution.   Note that $54 = 2 \times 3 \times 3 \times 3$. The only possibility for the probabilities that a white ball is drawn from each bag is 0.6 and 0.9. (Other attempted probabilities, such as 0.2 and 2.7 are inadmissible.) Therefore the probability that both balls are black is $(1 - 0.6)(1 - 0.9) = 0.04$. ♠

Comment.   The answer does not depend on the total number of balls being 25. We require only that the number of balls in the bag be such that the probabilities 0.6 and 0.9 are possible. For example, one bag could contain 5 balls of which 3 are white and the other 10 balls of which 9 are white. ♣

   While the answer is correct, the reader's approach neglects the possibility of having a factor in the numerator of the fraction representing one probability cancelled out by a factor in the denominator of the fraction representing the second probability. For example, the two probabilities for a white ball could be $7/10$ and $27/35$, in which case the probability of drawing two black balls is $(3/10) \times (8/35) = 12/175$. This is not equal to 0.04. If we ease the restriction that balls of both colors be in each urn, we could have one urn contain 50 balls of which 27 are white and the other any number of balls, all white. Then the probability of two black balls is zero.

8. An encounter in the cafeteria

The following problem was posed in the *Parade* column *Ask Marilyn* of February 19, 1995:

> If one couple eats lunch at a cafeteria twice a week (the day of the week varies), and they see another couple about 75% of the time, is there a logical reason for the first couple to assume that the second couple eats there more often that the first couple does?

Marilyn's answer is yes. Her reasoning is summarized in *CHANCE News* 4.04 (16 February to 3 March, 1995). Assuming a seven-day week, if both couples eat once a week, their chances of meeting are one in seven. If the second couple now goes twice a week, the chances of the first couple's seeing them double to two in seven. If, now, the first couple goes twice a week their chances of seeing the second couple double to four in seven. Since they in fact see the second couple more often than that, one can infer that the second couple dines there more frequently. Extending the argument, we find that, if one couple steps up the frequency to three times a week, then the probability of the first couple's seeing the second increases to 8/7.

To sort this out, suppose that couple A dines in the cafeteria a times per seven-day week. What is the probability that they will encounter couple B who eats there b times a week? If $a + b > 7$, then the couples will surely meet. Otherwise, couple B has $\binom{7}{b}$ choices of day, and $\binom{7-a}{b}$ of these choices will result in avoiding couple A. The probability that couple B will select a day in common with couple A is given by the following table:

$a \backslash b$	0	1	2	3	4	5	6	7
0	0	0	0	0	0	0	0	0
1	0	$\frac{1}{7}$	$\frac{2}{7}$	$\frac{3}{7}$	$\frac{4}{7}$	$\frac{5}{7}$	$\frac{6}{7}$	1
2	0	$\frac{2}{7}$	$\frac{11}{21}$	$\frac{5}{7}$	$\frac{6}{7}$	$\frac{20}{21}$	1	1
3	0	$\frac{3}{7}$	$\frac{5}{7}$	$\frac{31}{35}$	$\frac{34}{35}$	1	1	1
4	0	$\frac{4}{7}$	$\frac{6}{7}$	$\frac{34}{35}$	1	1	1	1
5	0	$\frac{5}{7}$	$\frac{20}{21}$	1	1	1	1	1
6	0	$\frac{6}{7}$	1	1	1	1	1	1
7	0	1	1	1	1	1	1	1

If over a period of time, the first couple, eating twice a week in the cafeteria, encounters the second couple 75 percent of the time, then it does seem likely from the foregoing table that the second eats there more than twice a week. If we interpret 75 percent as meeting during three weeks out of four, and if each couple eats twice a week in the cafeteria, then the probability that they will meet during at least three weeks out of four is $4 \cdot (11/21)^3 \cdot (10/21) + (11/21)^4 = 0.34$, which renders the event reasonably unlikely.

Column contributed by Elliot A. Weinstein of Baltimore, MD.

9. The car and goats and other problems

A problem which has recently become notorious is that of the "car and goats"; it is also known as Marilyn's problem or the Monty Hall problem. It was treated (along with problems **P**, **F**, and **C** below) by Sam C. Saunders in the April, 1990 newsletter *Mathematical Notes* (Vol. 33, #2; whole number 129) issued by the mathematics department of Washington State University in Pullman, WA. Almost a year and a half later, Marilyn Vos Savant published it in her *Parade* column, "Ask Marilyn" of September 8, 1991, unleashing a great controversy and an avalanche of articles in the public and professional press. In response, the *College Mathematics Journal* (24 (1993) 149–154; 26 (1995) 132–134; 27 (1996) 46, 205; 28 (1997) 44; 29 (1998) 136) tried to keep its readers up to date on developments.

The problem of the car and goats is related to other problems that are listed below. Several of them have been around for a long time and at least one goes back to the nineteenth century. As these problems have been amply treated in the literature, I will simply list some significant or unusual references; less accessible or more ephemeral ones can be found in the *College Mathematics Journal*.

Here are the problems:

M: *The problem of the car and goats.* A contestant in a game show is given a choice of three doors. Behind one is a car; behind each of the other two, a goat. She selects Door A. However, before the door is open, the host opens Door C and reveals a goat. He then asks the contestant: "Do you want to switch your choice to Door B?" Is it to the advantage of the contestant (who wants the car) to switch?

S: *The shell game.* A confidence man places a pea under one of three shells, out of sight of his mark. He then asks the mark to select the shell with the pea. After the mark picks a shell, the confidence man turns over one of the remaining shells, revealing no pea. He then asks the mark if he wishes to change his choice. Should the mark do so?

P: *The prisoner paradox.* Two of three prisoners are to be executed, but none of the prisoners knows which. One, A say, asks a guard: "Which of the other two is going to be executed? One of them will be and you will be giving me no information by telling me his name." The guard agrees and tells him that C is to be executed. A now thinks: "Before the guard said anything, my chances of being executed were 2 in 3. Now that I know it is either B or me, my chances are 1 in 2." Thus, the guard really has given information.

B: *Bertrand box problem.* Each of three boxes has two drawers. Each drawer of the first has a gold coin; each drawer of the second has a silver coin; the third box has a gold coin in one drawer and a silver one in the other. A box is chosen at random and a drawer opened to reveal a gold coin. What is the probability that the coin in the other drawer is silver?

C: *Three cards problem.* In a hat are three cards. Both sides of one are black; both sides of a second are red; one side of the third is black, while the other side is red. One card selected at random, is placed on the table. A red side is showing. What is the probability that the other side is black?

F: *Second sibling problem.* A family has two children, at least one of which is a boy. What is the probability that one is a girl? Does the answer change if it is given that the *elder* child is a boy?

A: *Paradox of the second ace.* What is the probability that a hand of two cards dealt from a four-card deck consisting of the aces of hearts and spades and the jacks of hearts and spades contains two aces given that it contains (a) at least one ace; (b) the ace of spades?

R: *Restricted choice.* You are South and declarer in a hand of bridge. North (dummy) holds ♠K 10 × × while you hold ♠A × × × ×. The remaining spades ♠Q J × × are in the other two hands. A low spade is led from the North hand; East produces the Queen which you take with the Ace, West following with a low spade. It is now your lead. Should you lead towards the King in the hope that one opponent has the singleton Jack, or should you expect West to have ♠J × and plan to finesse (i.e., play the Ten from dummy if West plays low)?

The standard analysis of problem **M** is based on the assumption that after the contestant makes the first choice, the host will always open an unselected door and reveal a goat (choosing the door randomly if both conceal goats) and then always offer the contestant the opportunity to switch. The contestant initially selects a door concealing a goat with probability 2/3. With a policy of always switching, she will win a car with this probability.

We will list the references by journal and indicate by a bold-face letter which of the problems is treated.

American Mathematical Monthly

Timothy Y. Chow, The Surprise Examination or Unexpected Hanging Paradox. 105 (1998) 41–51. **P**

Leonard Gillman, The Car and the Goats. 99 (1992) 3–7. **MAR**

American Statistician

William Bell, M. Bhaskara Rao (independently), Comments. 46 (1992) 241–242. **M**

Donald C. Butler, Letter to the editor. 20:5 (Dec., 1966) 36–37. **A**

John E. Freund, Puzzle or paradox? 19:4 (Oct., 1965) 29, 44. **A**

N.T. Gridgeman, Letter to the editor. 21:3 (June, 1967) 38–39. **A**

Ralph Haertel, Letter to the editor. 20:1 (Feb., 1966) 34–36. **A**

J.P. Morgan, N.R. Chaganty, R.C. Dahiya & M.J. Doviak, Let's make a deal (Comment by Richard G. Seymann) 45 (1991) 284–289. **MB**

Steve Selvin, A problem in probability (Letter to the editor). 29 (1975) 67. **M**

Steve Selvin, On the Monty Hall problem (Letter to the editor). 29 (1975) 134.

H. Fairfield Smith, "Puzzle or paradox?" – and Bayes (Letter to the editor). 21:2 (1967) 42–44. **A**

Chance

E. Engel and A. Venetoulias, Monty Hall's probability puzzle 4:2 (1991) 6–9.

Cognition

Maya Bar-Hillel & Ruma Falk, Some teasers concerning conditional probabilities 11 (1982) 109–122. **PCFA**

Ruma Falk, A closer look at the probabilities of the notorious three prisoners 43 (1992) 197–223.

Mathematical Gazette

Anthony Lo Bello, Ask Marilyn: the mathematical controversy in *Parade Magazine* 75 (1991) 275–277. **M**

Mathematical Scientist

V.V. Rao and M.B. Rao, A three-door game show and some of its variants 17 (1992) 89–94.

Mathematics Magazine

Steven J. Brams and Marc D. Kilgour, The box problem: To switch or not to switch 68 (1995) 27–34.

Mathematics Teacher

Robert Frankel, Monty's return (Reader reflection). 85 (1992) 176. **M**

Deborah Hecht, Simple dilemma (Reader reflection). 85 (1992) 90–91. **M**

Jeremy Kahan, Alternative solution for Monty (Reader reflection). 85 (1992) 90. **M**

S. Knight, Let's make a deal (Reader reflection). 85 (1992) 250, 252. **M**

Janet S. Milton & David L. Albig, To switch or ... (Reader reflection). 85 (1992) 8, 10. **M**

Norton Starr, A paradox in probability theory. 66 (1973) 166–168.

Personality and Social Psychology Bulletin

D. Gilovich, V.H. Medvec and S. Chen, Commission, omission , and dissonance reduction: Coping with regret in the "Monty Hall' problem. 21 (1995) 182–190.

D. Granberg and T.A. Brown, The Monty Hall dilemma. 21 (1995) 711–723.

Quantum

J.P. Georges and T.V. Craine, Generalizing Monty's dilemma. 5:4 (March/April, 1995) 17-21, 59–60.

Scientific American

Martin Gardner 196 (#4, April, 1957) 166 [Reprinted in *Mathematical Puzzles and Diversions,* Simon & Schuster, 1959, pages 49–51.] **A**

Martin Gardner 200 (#5, May, 1959) 166; 200 (#6, June, 1959) 164; 201 (#4, Oct., 1959) 180, 182; 201 (#5, Nov., 1959) 188. [Reprinted in *Second Scientific American Book of Mathematical Puzzles and Diversions,* Simon & Schuster, 1961.] **PFA**

J. Michael Shaughnessy & Thomas Dick, Monty's delimma: should you stick or switch? 84 (1991) 252–256. **M**

Skeptical Inquirer

Kendrick Frazier, 'Three door' problem provokes letters, controversy; John Geohegan, SI readers show their stuff (follow-up and letters) 16 (1992) 192–199. **M**

Martin Gardner, Probability paradoxes (Notes of a fringe watcher) 16 (1992) 129–132. **SPFA**

Gary P. Posner, Nation's mathematicians guilty of 'innumeracy'. 15 (1991) 342–345. **M**

Letters Column. 16 (1992) 440–443. **PFM**

Books

W.W. Rouse Ball & H.S.M. Coxeter, *Mathematical Recreations and Essays,* 12th edition, University of Toronto, 1974, page 44. **A**

J. Bertrand, *Calcul des Probabilités* Gauthier-Villars, 1888, page 2. **B**

Stephen J. Campbell, *Flaws and Fallacies in Statistical Thinking,* Prentice-Hall, 1974, page 132. **C**

F. Mosteller, *Fifty Challenging Problems in Probability with Solutions,* Addison-Wesley, 1965, Problem 13, page 28. **P**

Anatol Rapoport, *Decision Theory and Decision Behaviour,* Kluwer, 1989, pages 72–75. **PF**

Gábor J. Székely, *Paradoxes in Probability Theory and Mathematical Statistics,* Akademiai Kiadó, Budapest; D. Reidel, Dordrecht, Holland, 1986, pages 68–69. **P**

Marilyn Vos Savant, *The Power of Logical Thinking: easy lessons in the art of reasoning ... and hard facts about its absence in our lives,* St. Martin's Press, New York, 1996, appendix, pages 169–196. **M**

IO. Your lucky number is in Pi

In a regional magazine for high school teachers, the following passage appears:

> There is only one place where you can find next week's winning Lotto 6-49 numbers. .... Next week's winning numbers are located in the decimal expansion of π. The decimal expansion of π is an infinite sequence of random numbers. Therefore, any number (as many digits long as you wish) must appear in the sequence with a probability of one! In other words, if next week's winning numbers are 12, 15, 29, 42, 46, and 48, then you can be assured that the decimal expansion of π can be written as $3.14 \ldots 121529424648 \ldots$.

This passage might generate some interesting discussion in a probability and statistics class. Are the digits of π random? What can possibly be meant by the assertion that "π is an infinite sequence of random numbers"? Do you have any idea what the background of the statement is? Granting some interpretation to the author's assertions, does it necessarily follow that any particular finite succession of digits must occur in the decimal expansion of π?

Chapter 6

CALCULUS: LIMITS AND DERIVATIVES

I. All powers of x are constant.

Proposition. Let n be a nonnegative integer. The function x^n is constant.

Proof. Observe that $(x^0)' = 0$. Assume that the derivative of x^n is zero for $n = 0, 1, 2, \ldots, k$. Then

$$(x^{k+1})' = (x \cdot x^k)' = x' \cdot x^k + x \cdot (x^k)'$$

is also zero since $x' = (x^1)' = (x^k)' = 0$. ♡

   Contributed by Alex Kuperman of the Israel Institute of Technology (Technion) in Haifa.

2. Differentiating the square function

At $x = c$, the function $y = (x - c)^2 = x^2 - 2cx + c^2$ has a minimum, so that $0 = Dy = D(x^2) - 2cD(x) = D(x^2) - 2c$. But c is arbitrary and $c = x$. Hence $D(x^2) = 2c = 2x$.

   Contributed by A.W. Walker of Toronto, ON.

3. 3 equals 2

Let x be positive. Differentiating the equation $x^3 = x^2 + x^2 + \cdots + x^2$ (to x terms) yields $3x^2 = 2x + 2x + \cdots + 2x = x(2x) = 2x^2$, whence $3 = 2$. ◊

   An alternative proof of the same fact goes like this. Let x be constant with the value 1. Then $x = x^2 = x^3$. Now set $y = x$. Then $y = x^2$ and $y = x^3$. Therefore $dy/dx = 2x$ and $dy/dx = 3x^2$. Therefore dy/dx is both constant with value 2 and constant with value 3. ◊

In the second argument, the trouble seems to be that we are differentiating with respect to a constant. Is there anything in the definition of dy/dx that prevents x from being constant? In fact, the naive definition does not ensure that dy/dx is uniquely determined by x and y. The following definition will ensure this: if there is a function f, whose domain is the set of values of x such that $y = f(x)$, then $dy/dx = f'(x)$. The restriction on the domain ensures that f is uniquely determined by x. And if x is constant, the domain of f is a single number and so f is not differentiable and dy/dx does not exist.

There is a flavor here of treating a constant as a nascent variable, and my mind was drawn to two considerations which may be related to the issue at hand: the values of $1 - 1 + 1 - 1 + \cdots$ and the distinction between the δy and dy operations in the calculus of variations. In the case of $1 - 1 + 1 - 1 + \cdots$, the constant 1 can be treated as the limiting value of a variable and the series evaluated using the Leibnizian principle that "what is true up to the limit is true at the limit". However, as Callet observed in the latter part of the 18th century, the answer depends on the choice of variable. For, if $m < n$,

$$\frac{1 + x + \cdots + x^{m-1}}{1 + x + \cdots + x^{n-1}} = \frac{1 - x^m}{1 - x^n} = 1 - x^m + x^n - x^{n+m} + x^{2n} - \cdots.$$

Setting $x = 1$ yields $m/n = 1 - 1 + 1 - 1 + 1 - \cdots$.

As for the "calculus of variations" approach, let us perturb the constant 1 evaluated at one place by a delta amount t and at another by a delta amount u. Then the differential quotients for 1^2 and 1^3 become

$$\frac{(1+t)^2 - (1+u)^2}{(1+t) - (1+u)} = 2 + (t + u)$$
$$\frac{(1+t)^3 - (1+u)^3}{(1+t) - (1+u)} = 3 + 3(t + u) + (t^2 + tu + u^2).$$

Suppressing the perturbations gives the result.

The first argument due to R. L. Francis of Southeast Missouri State University, the second due to Hugh Thurston of the University of British Columbia in Vancouver.

4. The shortest distance from a point to a parabola

Problem. Determine the shortest distance from the point $(0, 5)$ to a parabola $16y = x^2$.

Solution. We must minimize $f(y) = x^2 + (y-5)^2 = 16y + (y-5)^2$. Since $f'(y) = 2y + 6$, the only critical value of f is $y = -3$, which corresponds to an imaginary value of x. Hence the minimum distance does not exist. ♠

Observe that $f(y)$ is defined only for nonnegative values of y, since y is the ordinate of a point on the curve $16y = x^2$. Since the circle $x^2 + (y-5)^2 = 5^2$ meets the parabola only at $(0,0)$, the function is minimized for $y = 0$. This is an endpoint extremum which one cannot expect to pick up from the vanishing of the first derivative.

A similar problem is discussed in a number of articles reprinted in the book, *Selected Papers on Calculus,* (MAA, 1969):

C.S. Ogilvy, Exceptional extremum problems, *Amer. Math. Monthly* 67 (1960) 270–275; S.P. 262–267.

Hugh A. Thurston, So-called "exceptional" extremum problems, *Amer. Math. Monthly* 68 (1961) 650–652; S.P. 268–270.

See also the papers:

C.O. Oakley, End-point maxima and minima, *Amer. Math. Monthly* 54 (1947) 407-409; S.P. 244–246.

Frank Hawthorne, A simple endpoint minimum, *Amer. Math. Monthly* 58 (1951) 188; S.P. 256.

Fallacious solution contributed by David M. Bloom of Brooklyn College, CUNY, New York, NY.

5. A foot by any other name

Suppose that we have a stick one foot long leaning against a wall, as shown in Figure 6.5. We want to find the angle θ that will make the quantity $x^2 + y$

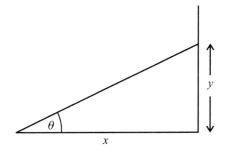

FIGURE 6.5

as large as possible. Since $x = \cos\theta$ and $y = \sin\theta$, we have to maximize $\cos^2\theta + \sin\theta$ with respect to θ. We obtain $\theta = \pi/6$.

Now repeat the calculation assuming that the stick is twelve inches long. Since $x = 12\cos\theta$ and $y = 12\sin\theta$, we are maximizing $144\cos^2\theta + 12\sin\theta$. The answer is now given by $\sin\theta = 1/24$, so $\theta \neq \pi/6$. We conclude that one foot does not equal twelve inches. ♣

The difficulty is with the inconsistent dimensions of the terms of the expression $x^2 + y$. To remedy the situation, we should make the expression dimensionally homogeneous by providing y with a coefficient R, which we can think of as representing the number of units per foot. Thus $R = 1$ when the measurements are in feet and $R = 12$ when the measurements are in inches. We find that the answers are consistent.

Inconsistency contributed by David Protas of the California State University at Northridge and comment due to Joel M. Simon of Central High School in Philadelphia, PA.

6. A degree of differentiation

Let x be the measure of an angle in radians. Then the corresponding measure of the angle in degrees is $y = 360x/2\pi = \alpha x$, where we set $\alpha = 180/\pi$ for convenience. Now

$$\frac{d}{dy}\sin y = \frac{d}{dx}\sin\alpha x \cdot \frac{dx}{dy} = \alpha\cos\alpha x \cdot \frac{1}{\alpha} = \cos y,$$

for any x in radians and y in degrees. Now imagine that x is a function of t, such that $dx/dt = V \neq 0$ for all t. Thus $dy/dt = \alpha V$. Now

$$\frac{d}{dt}\sin x = \cos x \cdot \frac{dx}{dt} = V\cos x$$

and

$$\frac{d}{dt}\sin y = \cos y \cdot \frac{dy}{dt} = \alpha V\cos y.$$

When y in degrees equals αx with x in radians, then $\sin y = \sin x$, so the derivatives just computed must be equal. In particular, when $y = x = 0$, we get $V = \alpha V$, so that $\alpha = 1$. ♣

Note that $\sin y$ with y in degrees is really a different function than $\sin x$ with x in radians. Using $\sin x$ for the latter, we find that the sine function for y in degrees is $\sin(y/\alpha)$, which we denote by $S(y)$. Then

$$\frac{d}{dy}S(y) = \frac{d}{dy}\sin(y/\alpha) = \frac{1}{\alpha}\cos(y/\alpha) = \frac{1}{\alpha}C(y)$$

where $C(y)$ is the cosine function for y in degrees. Carrying through this clarification eliminates the contradiction.

Students who are confused should be encouraged to sketch on the same axes the graphs of $\sin x$ and $S(x)$, and of $\cos x$ and $C(x)$. They should note that the curves are related by a horizontal dilatation, which accounts for the factor arising in the derivative of $S(x)$.

Contributed by David Singmaster of the South Bank Polytechnic in London, England, who originally published it in Mathematical Spectrum 13:3 (1980/81) 76, 83.

7. The derivative of the sum is the sum of the derivatives

It is readily established that

$$\sum_{k=1}^{n}(2k-1)^2 = \frac{4n^3 - n}{3}.$$

With this result in hand, we can determine $s(n) = \sum_{k=1}^{n}(2k-1)^3$. Observe that

$$\frac{ds(n)}{dn} = \sum_{k=1}^{n}\frac{d}{dk}(2k-1)^3 = 6\sum_{k=1}^{n}(2k-1)^2 = 8n^3 - 2n,$$

whence $s(n) = 2n^4 - n^2 + C$. Since $s(1) = 1$, we must have that $C = 0$, so $s(n) = n^2(2n^2 - 1)$.

8. Differentiating x^x

How should we differentiate x^x? If the base were constant, we should get $x \cdot x^{x-1} = x^x$. On the other hand, if the exponent were constant, we should get $(\log x)x^x$. But neither is constant, so we should use both terms to get

$$D(x^x) = x^x + (\log x)x^x = (1 + \log x)x^x.$$

This method works for differentiating any function of the form $f(x)^{g(x)}$, as can be easily justified using the chain rule. James F. Hurley makes this point in his books, *Calculus* (Wadsworth, 1987, Section 6.4, p. 364) and *Calculus, A Contemporary Approach* (McGraw-Hill, 1992). See also the notes

Fred Halpern, Using the multivariate chain rule, *Amer. Math. Monthly* 92 (1985) 144–145.

Mark Galit, Simplifying logarithmic differentiation, *AMATYC Review* 6 (#1, 1984) 29–30.

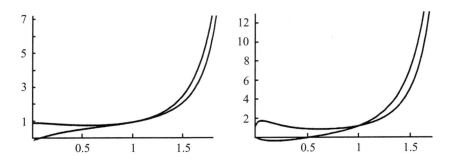

FIGURE 6.9

9. Double exponential

Because exponentiation is non-associative, the expression $y = x^{x^x}$ can be interpreted in two different ways. When $y = (x^x)^x = x^{(x^2)}$, then $y' = x^{x^x}(2x \log x + x)$, while if $y = x^{(x^x)}$, then $y' = x^{x^x}(x^x(\log x + 1) \log x + x^{x-1})$. The two graphs for these interpretations are shown in Figure 6.9. Which of these will your software package give?

   Diagrams provided by Leszek Gawarecki of the GMI Engineering and Management Institute in Flint, MI.

IO. Iterated exponential

Problem. Solve

$$x^{x^{x^{\cdots}}} = 2.$$

Solution. Since the exponent of x is equal to 2, we find that $x^2 = 2$ and so $x = \sqrt{2}$. ♡

Problem. Solve

$$y^{y^{y^{\cdots}}} = 4.$$

Solution. Since the exponent of y is equal to 4, we find that $y^4 = 4$ and so $y = \sqrt{2}$. ♠

   But these solutions raise the question: what, really, is

$$\sqrt{2}^{\sqrt{2}^{\sqrt{2}^{\cdots}}} ?$$

To analyze this anomalous situation, we need to decide what is meant by an infinite exponential

$$v = u^{u^{u^{u^{\cdots}}}}.$$

It is natural to define it as the limit of the sequence $\{u, u^u, u^{u^u}, \ldots\}$, i.e.,

$$v = \lim_{n \to \infty} u_n,$$

where $u_1 = u$ and $u_n = u^{u_{n-1}}$ for $n \geq 2$. Letting n tend to infinity, we find that v must satisfy $u^v = v$, so that v is a fixed point of the function u^x.

If $u^x = x$, then $\log u$ has the form $(\log x)/x$, a function that assumes its maximum value of $1/e$ when $x = e$. When $u > e^{1/e}$, then $u^x > x$ and the sequence $\{u_n\}$ increases without bound.

If $1 < u < e^{1/e}$, we have the graph in Figure 6.10. Thus u^x has two fixed points, v, w with $v < 1/e < w$. The lesser v is attracting and the greater w repelling. Any sequence $\{x_n\}$ with $1 < x_1 < w$ and $x_n = u^{x_{n-1}}$ for $n \geq 2$ must have limit v. In particular, since $u^u > u^1 = u$, we have $1 < u < v$ and the sequence defined above increases to the limit v. In particular, the value of

$$\sqrt{2}^{\sqrt{2}^{\sqrt{2}^{\cdots}}}$$

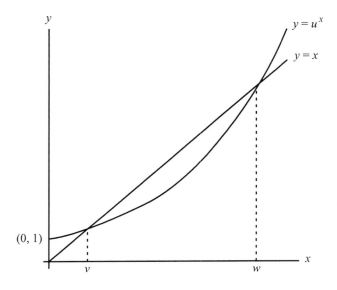

FIGURE 6.10

is 2. That it is not 4 is corroborated by the fact that the sequence is increasing and never exceeds 2.

Suppose that $0 < u < 1$. Then the even and odd terms of the sequence $\{u_n\}$ are respectively decreasing and increasing. Let t be defined by the equation $u^{u^t} = t$, or equivalently $-\log(1/u) = (1/u)^t \log t$. When $0 < u < e^{-e}$, then this equation has three solutions, the smallest and largest of which are the limits of the alternate entries of the sequence $\{u_n\}$; when $e^{-e} \le u < 1$, the solution is unique.

Then the domain of the function v is the set $\{u : e^{-e} \le u \le e^{1/e}\}$. This function has been treated in several places in the literature:

D. F. Barrow, Infinite exponentials, *Amer. Math. Monthy* 43 (1936) 150–160.

Ted Courant, Towers of powers: a potent paradox, *Mathematical Journal* 3 (1993) 60–64.

R. A. Knoebel, Exponentials reiterated, *Amer. Math. Monthly* 88 (1981) 235–252.

M.C. Mitchelmore, A matter of definition, *Amer. Math. Monthly* 81 (1974) 643–647.

Ivan Niven, *Maxima and minima without calculus*, MAA, 1981 (see problem M1 and its discussion, pages 242, 291–292).

P. J. Rippon, Infinite exponentials, *Math. Gazette* 67 (1983) 189–196.

G. T. Vickers, More about an infinite exponential, *Mathematical Spectrum* 27:3 (1994–1995), 54–56. [Reviewed in *College Math. J.* 27 (1996) 159.]

Knoebel has a comprehensive bibliography that should be consulted by anyone wishing to add to the discussion. Rippon gives a treatment of an infinite exponential equation that is accessible to high school students.

Comments and references contributed by Ivan Niven of Eugene, OR; Man-Keung Siu of the University of Hong Kong. Herman Sutton of School-craft College in Livonia, MI; Robert E. Terrell of Cornell University in Ithaca, NY; Brian Stewart Watts of Roanoke Rapids, NC; Donald F. Young of Southern College of Technology in Marietta, GA.

II. Calculation of a limit

Suppose that $L = \lim_{x \to 0}(\sin 3x)/(x^3)$. Then

$$L = \lim_{x \to 0} \frac{3\sin x - 4\sin^3 x}{x^3} = 3\lim_{x \to 0} \frac{\sin x}{x^3} - 4\lim_{x \to 0}\left(\frac{\sin x}{x}\right)^3$$

$$= 3\lim_{t \to 0}\frac{\sin(3t)}{(3t)^3} - 4 = \frac{1}{9}\lim_{t \to 0}\frac{\sin 3t}{t^3} - 4 = \frac{1}{9}L - 4,$$

whence $L = -9/2$. ♡

This can be confirmed by using l'Hôpital's Rule:

$$\lim_{x \to 0} \frac{\sin 3x}{x^3} = \lim_{x \to 0} \frac{\cos 3x}{x^2} = \frac{-3}{2} \lim_{x \to 0} \frac{\sin 3x}{x} = \frac{-3}{2}(3) = \frac{-9}{2},$$

or

$$L = \lim_{x \to 0} \frac{3 \sin x - 4 \sin^3 x}{x^3} = 3 \lim_{x \to 0} \frac{\sin x}{x^3} - 4 \lim_{x \to 0} \frac{\sin^3 x}{x^3} = \lim_{x \to 0} \frac{\cos x}{x^2} - 4$$

$$= -\lim_{x \to 0} \frac{\sin x}{2x} - 4 = -\frac{1}{2} - 4 = -\frac{9}{2}. \quad \diamond$$

Contributed by Cherie D'Mello and Joel Chan of the University of Toronto in Ontario and by John M. Cohen of Shawnee, KS.

12. Which is the correct asymptote?

Problem. Find the slant asymptote of the curve with equation

$$y = \frac{x^2 + 3x + 7}{x + 2}.$$

First solution. By division, we find that

$$\frac{x^2 + 3x + 7}{x + 2} = x + 1 + \frac{5}{x + 2}.$$

Since the final term tends to zero as x grows, the asymptote is the line of equation $y = x + 1$.

Second solution. Following a procedure frequently used in calculating limits at infinity, we find that

$$\frac{x^2 + 3x + 7}{x + 2} = \frac{x + 3 + 7/x}{1 + 2/x}.$$

For large x, the value is approximately $x + 3$, so the asymptote should be the line of equation $y = x + 3$. ♠

Ragnar Dybvik of Tingvoll, Norway, points out that Solution 2 can be adapted to give the correct answer through the following manipulation:

$$\frac{x^2 + 3x + 7}{x + 2} = \frac{x + 3 + \dfrac{7}{x}}{1 + \dfrac{2}{x}} = \frac{\left(x + 3 + \dfrac{7}{x}\right)\left(1 - \dfrac{2}{x}\right)}{\left(1 + \dfrac{2}{x}\right)\left(1 - \dfrac{2}{x}\right)}$$

$$= \frac{x + 1 + \dfrac{1}{x} - \dfrac{14}{x^2}}{1 - \dfrac{4}{x^2}}.$$

Carl Crockett of DFMS, USAFA, CO, presented the argument leading to the two answers to three different classes with a variety of results. He found each experience interesting.

In a computer lab session I presented the two candidate asymptotes. I asked the students to determine which was correct. With the aid of computer graphics they were able to do so quite quickly. With answer in hand they showed little (no) interest in algebraic justification. I could not generate any concern for the validity of either approach.

My next presentation was made in a classroom with no computer. I presented the two candidates and asked for suggestions on how to determine which was correct. There were no useful responses. I suggested that both approaches, applied to the limiting value of the function, correctly established that the limit did not exist. I further suggested that it may be inappropriate to try and identify an asymptote from quantities which are best described by the phrase "does not exist." No discussion followed.

The very next hour, with a different section of students, the same question generated lively discussion. The best, and most leading question was, "What if I didn't have any candidates? Then how would I find the asymptote?" This allowed me to discuss the idea of what an asymptote is and how we would look for one if we were to hypothesize the existence of a linear one. In particular, we would search for a and b such that

$$\lim_{x \to \infty} \left| (ax + b) - \frac{x^2 + 3x + 7}{x + 2} \right| = 0.$$

We carried out the algebra on the board and everyone felt increased confidence in their understanding of the notion of asymptote.

I was also able to successfully parlay the discussion into a promotion for the need to occasionally consider the theory underlying a particular problem.

Sandra Z. Keith of St. Cloud State University in Minnesota gave her class the task of finding slant asymptotes for the graph of $y = (x + 1)^2/(x - 1)$. The common approach was to write

$$y = \frac{x^2 + 2x + 1}{x - 1} = \frac{1 + \dfrac{2}{x} + \dfrac{1}{x^2}}{\dfrac{1}{x} - \dfrac{1}{x^2}}$$

and come to a halt. Other answers obtained were $y = x + 3$ by division, $y = x + 2$ by dividing numerator and denominator by x, and $y = x + 1$ by writing the function as $x + 1$ times $(x + 1)/(x - 1)$.

Problem and solutions contributed by Michael G. Murphy of the University of Houston (Downtown) in Texas.

13. Every derivative is continuous

Using the definition of derivative and then an application of l'Hopital's Rule, we obtain the following proof that the derivative is continuous at every point:

$$f'(a) = \lim_{x \to a} \frac{f(x) - f(a)}{x - a} = \lim_{x \to a} f'(x). \; \clubsuit$$

For each x close to a,

$$\frac{f(x) - f(a)}{x - a} = f'(u_x)$$

for some u_x lying between x and a. But the u_x that can be used may have to come from a sparse set of numbers close to a. For example, let $a = 0$ and $f(x) = x^2 \sin(1/x)$ when $x \neq 0$ and $f(0) = 0$. Then $f'(0) = 0$ and, when x is small, $f(x)/x = x \sin(1/x)$ lies close to zero. Thus, while u_x and $f'(u_x)$ are close to zero with x, the derivative $f'(x) = 2x \sin(1/x) - \cos(1/x)$ whips over the whole of the interval $(-1, 1)$ as x approaches 0 and so $\lim_{x \to 0} f'(x)$ does not exist.

Contributed by F. V. Rickey of Bowling Green State University and E. Merkes of the University of Cincinnati, both in Ohio.

14. Telescoping series

Consider the sum

$$\sum_{k=1}^{\infty} \frac{1}{(k+1)(k+2)}.$$

On the one hand,

$$\frac{1}{6} + \frac{1}{12} + \frac{1}{20} + \cdots = \left(\frac{1}{2} - \frac{1}{3}\right) + \left(\frac{1}{3} - \frac{1}{4}\right) + \left(\frac{1}{4} - \frac{1}{5}\right) + \cdots = \frac{1}{2};$$

while on the other,

$$\frac{1}{6} + \frac{1}{12} + \frac{1}{20} + \cdots = \left(1 - \frac{5}{6}\right) + \left(\frac{5}{6} - \frac{3}{4}\right) + \left(\frac{3}{4} - \frac{7}{10}\right) + \left(\frac{7}{10} - \frac{2}{3}\right) + \cdots = 1. \; \clubsuit$$

In the first rendition, the partial sum of the first n terms is

$$\sum_{k=1}^{n} \left(\frac{1}{k+1} - \frac{1}{k+2}\right) = \frac{1}{2} - \frac{1}{n+2},$$

and its limit is clearly $1/2$. However,

$$\sum_{k=1}^{n} \frac{1}{(k+1)(k+2)} = \frac{1}{2} \sum_{k=1}^{n} \left(\frac{k+3}{k+1} - \frac{k+4}{k+2}\right)$$

$$= \frac{1}{2}\left[2 - \frac{n+4}{n+2}\right] = 1 - \frac{n+4}{2(n+2)}.$$

The term $(n+4)/(2(n+2))$ cannot be neglected as n tends to infinity, and, as before, the nth partial sum is seen to converge to $\frac{1}{2}$ rather than 1.

While a freshman at Connecticut College in New London, Eleanor A. Maddock analyzed this in a class paper.

Chapter **7**

CALCULUS: INTEGRATION AND DIFFERENTIAL EQUATIONS

I. A new way to obtain the logarithm

Exercise. Integrate $\int \frac{1}{x+1} dx$.

Solution (by a student on a quiz).

$$\int \frac{1}{x+1} = \int \left(\frac{1}{x} + \frac{1}{1} \right) dx = \int \frac{1}{x} dx + \int \frac{1}{1} dx = \log x + \log 1$$
$$= \log(x+1) + C. \quad \Diamond$$

Contributed by Lewis Lum of the University of Portland in Oregon.

2. The constant of integration

A fair number of fallacies can be generated by simply neglecting the constant of integration when computing indefinite integrals. Here are a few examples.

(a) In integrating $J = \int dx/(x \log x)$ by parts with $u = 1/\log x$ and $dv = dx/x$, one readily obtains $J = 1 + J$. Cancelling J yields $1 = 0$.

According to R. P. Boas of Seattle, WA, all swindles like this reduce to the form $\int (g'/g) dx$ and he mentions the examples $\int e^x e^{-x} dx$ and the "perfectly reasonable" looking

$$\int \frac{4 \cos^2 x \, dx}{2x + \sin 2x}.$$

He believes that the earliest occurrence of this fallacy (for $\int dx/x$) was J. L. Walsh, A paradox resulting from integration by parts, *American Mathematical Monthly* 34 (1927) 88. Other appearances are in E. A. Maxwell, *Fallacies in mathematics* (Cambridge, 1959) and *Mathematical Gazette* 57 (1973) 200–201; Note 287. In the latter case, the integral considered is $\int \tan x \, dx$.

(b) Integrating by parts twice yields

$$\int e^x \sinh x \, dx = e^x \cosh x - \int e^x \cosh x \, dx$$

$$= e^x \cosh x - \left\{ e^x \sinh x - \int e^x \sinh x \, dx \right\}$$

whence

$$e^x (\cosh x - \sinh x) = 0.$$

An immediate corollary is that $\cosh x = \sinh x$ and $1 = e^x e^{-x} = 0$. However, if one is more careful and inserts an arbitrary constant with each integration by parts, one readily concludes that $c = e^x (\cosh x - \sinh x) = 0$ for all real c.

For other examples of the baleful effect of neglecting the integration constant, consult Phillip J. Sloan, The significance of the "insignificant constants" (Sharing teaching ideas) in *Mathematics Teacher* 82 (1989) 186, 188.

(a) and (b) contributed by Bernard C. Anderson of Allen Park, MI and Robert Weinstock of the Oberlin State College Physics Department in Oberlin, OH respectively.

3. The integral of log sin x

Let $I = \int \log(\sin x) dx$. Then

$$I = \int \log \left(2 \sin \frac{x}{2} \cos \frac{x}{2} \right) dx$$

$$= \int (\log 2) dx + \int \log \left(\sin \frac{x}{2} \right) dx + \int \log \left(\cos \frac{x}{2} \right) dx$$

$$= (\log 2)x + \int \log \left(\sin \frac{x}{2} \right) dx + \int \log \left(\cos \frac{x}{2} \right) dx.$$

In the last two integrals, let $u = x/2$. Then

$$I = (\log 2)x + 2 \int \log(\sin u) du + 2 \int \log(\cos u) du.$$

Since the integration variable is dummy, the identity becomes

$$I = (\log 2)x + 2I + 2 \int \log(\cos u) du,$$

and so

$$I = -(\log 2)x - 2\int \log(\cos u)du$$

$$= -(\log 2)x - 2\int \log\left(\sin\left(\frac{\pi}{2} - u\right)\right)du.$$

Letting $t = \pi/2 - u$ yields

$$I = -(\log 2)x + 2\int \log(\sin t)dt = -(\log 2)x + 2I.$$

This is equivalent to

$$\int \log(\sin x)dx = (\log 2)x + C$$

where C is an arbitrary constant. $\diamond$

Contributed by Russ Euler of Northwest Missouri State University in Marysville.

4. Evaluation of a sum

Problem.   Evaluate

$$\sum_{n=1}^{\infty} \frac{1}{(2n-1)(3n-1)}.$$

Solution by Joe Howard, New Mexico Highlands University, Las Vegas, NM.

$$\sum_{n=1}^{\infty} \frac{1}{(2n-1)(3n-1)} = \sum_{n=1}^{\infty}\left(\frac{2}{2n-1} - \frac{3}{3n-1}\right)$$

$$= \frac{2}{1} - \frac{3}{2} + \frac{2}{3} - \frac{3}{5} + \frac{2}{5} - \frac{3}{8} + \cdots.$$

Let

$$f(x) = \frac{2x}{1} - \frac{3x^2}{2} + \frac{2x^3}{3} - \frac{3x^5}{5} + \cdots$$

$$= 2\left(\frac{x}{1} + \frac{x^3}{3} + \frac{x^5}{5} + \cdots\right) - 3\left(\frac{x^2}{2} + \frac{x^5}{5} + \frac{x^8}{8} + \cdots\right)$$

for $0 \le x \le 1$. The series is absolutely convergent for $|x| < 1$, so the terms can be rearranged. It follows that

$$f'(x) = 2(1 + x^2 + x^4 + \cdots) - 3(x + x^4 + x^7 + \cdots)$$

$$= 2\left(\frac{1}{1-x^2}\right) - 3x\left(\frac{1}{1-x^3}\right) = \frac{x+2}{(x+1)(x^2+x+1)}$$

$$= \frac{1}{x+1} + \frac{-x+1}{x^2+x+1}.$$

Thus,

$$f(x) = \int_0^x \frac{dt}{t+1} - \frac{1}{2}\int_0^x \frac{(2t+1)dt}{t^2+t+1} + \frac{3}{2}\int_0^x \frac{dt}{(t+\frac{1}{2})^2 + (\frac{\sqrt{3}}{2})^2}$$

$$= \log|x+1| - \frac{1}{2}\log|x^2+x+1|$$

$$+ \sqrt{3}\left(\arctan\frac{2x+1}{\sqrt{3}} - \arctan\frac{1}{\sqrt{3}}\right).$$

Since the series for $f(x)$ is convergent for $x = 1$, the given series converges to

$$f(1) = \log\frac{2}{\sqrt{3}} + \frac{\sqrt{3} \cdot \pi}{6}. \quad \heartsuit$$

This problem was posed in *Crux Mathematicorum* 22 (1996) 219; 23 (1997) 371–373, where the (correct) answer given was

$$2\log 2 - \frac{3}{2}\log 3 + \frac{\sqrt{3} \cdot \pi}{6}.$$

5. Integrals of products

Here are some examples in which the integral of the product is equal to the product of the integrals; it is an interesting undergraduate investigation to provide some general conditions to enable this.

$$\int \frac{x+2}{x^3}dx = \int (x+2)dx \cdot \int x^{-3}dx = \frac{(x+2)^2}{2} \cdot \frac{x^{-2}}{-2} + C$$

$$= -\frac{1}{4}\left(\frac{x+2}{x}\right)^2 + C.$$

$$\int e^{4x}dx = \int (e^{2x})^2 dx = \left(\int e^{2x}dx\right)^2$$

$$= \left(\frac{e^{2x}}{2}\right)^2 + C = \frac{1}{4}e^{4x} + C.$$

Contributed by W. Heierman of the University of Science and Technology in Kumasi, Ghana.

6. L'Hopital's Rule under the integral sign

Exercise. Evaluate $\int_1^\infty (x-1)e^{-x}dx$.

Solution. The integral is equal to

$$\int_1^\infty \frac{x-1}{e^x}dx = \int_1^\infty \frac{1}{e^x}dx = \frac{1}{e},$$

where l'Hopital's Rule is used to pass from the first to the second member.
♡

A student solution contributed by Peter Lindstrom of North Lake College in Irving, TX.

7. A power series representation

Problem. Expand about the origin $f(x) = (1+x^2)/(1-x^2)$.

Solution. By the quotient rule, we find that

$$f'(x) = 2\left[\frac{2x}{(1-x^2)^2}\right] = 2\left[\frac{1}{(1-x^2)}\right]',$$

whence $f(x) = 2(1-x^2)^{-1} = 2(1+x^2+x^4+x^6+\cdots)$ (so that, in particular, $f(0) = 2$). ♠

All we know about functions sharing the same derivative is that they differ by a constant. Here we can check directly that $f(x) = 2(1-x^2)^{-1} - 1$.

Contributed by Joe Howard of the New Mexico Highlands University, Las Vegas.

8. More fun with series

In the following, recall that Abel's Theorem provides that if $\sum a_n$ converges and $f(x) = \sum a_n x^n$ is defined for $|x| < 1$, then $\lim f(x) = \sum a_n$ when x approaches 1 from below.

Observe that

$$\frac{1}{1+t} = \frac{1}{1-t^2} - t\frac{1}{1-t^4} - t^3\frac{1}{1-t^4}$$
$$= 1 - t - t^3 + t^2 - t^5 - t^7 + t^4 - t^9 - t^{11} + \cdots$$

for $|t| < 1$. Integrating both sides between 0 and x yields

$$\log(1+x) = x - \frac{x^2}{2} - \frac{x^4}{4} + \frac{x^3}{3} - \frac{x^6}{6} - \frac{x^8}{8} + \cdots \qquad (|x| < 1).$$

Taking the limit as x tends to 1 and invoking Abel's theorem, we obtain $\log 2 = 1 - \frac{1}{2} - \frac{1}{4} + \frac{1}{3} - \frac{1}{6} - \frac{1}{8} + \cdots$.

However, grouping terms in the series yields

$$\left(1 - \frac{1}{2}\right) - \frac{1}{4} + \left(\frac{1}{3} - \frac{1}{6}\right) - \frac{1}{8} + \cdots = \frac{1}{2} - \frac{1}{4} + \frac{1}{6} - \frac{1}{8} + \cdots$$

$$= \frac{1}{2}\left(1 - \frac{1}{2} + \frac{1}{3} - \frac{1}{4} + \cdots\right)$$

$$= \frac{1}{2}\log 2.$$

Thus, $\log 2 = \frac{1}{2}\log 2$. ♣

Annie Selden of the Tennessee Technological University in Cookeville invites us to consider an example in J. Stewart, *Calculus* (Brooks/Cole, 1987, pages 606–607). The author rearranges the harmonic series to the series given above, inserts the parentheses and obtains $\frac{1}{2}\log 2$. David Bloom of Brooklyn College, NY comments:

> The fallacy in Stewart's argument is entirely different from that in the FFF. Whereas Stewart, as you point out, "rearranges the [conditionally convergent] alternating harmonic series," the FFF rearranges the *absolutely* convergent series for $\log(1+x)$ (perfectly legal!) but then tries to take the limit termwise, appealing to Abel's theorem that the limit of a convergent power series equals the series of limits if the latter series converges (even only conditionally). What makes the argument invalid is that the series whose limit we are taking is *no longer a power series,* since its exponents no longer appear in increasing order; thus Abel's theorem in inapplicable.

The misapplication of Abel's Theorem contributed by Frank Burk of California State University in Chico.

9. Why integrate?

Note that

$$\int_{-1}^{2} x^2\,dx = [x^2]_{-1}^{2} = 3$$

and that

$$\int_{-1}^{2} (x+1)^2 dx = \left[(x+1)^2\right]_{-1}^{2} = 9. \diamond$$

Contributed by James C. Kirby of Tarleton State University, Stephenville, TX.

10. The disappearing factor

A standard problem in freshman calculus is to have the students verify that

$$\int_{0}^{\infty} \frac{x}{1+x^3} dx = \int_{0}^{\infty} \frac{1}{1+x^3} dx.$$

While simply removing the factor does not make the integral easier to compute in this case, there are other situations in which it is decidedly helpful:

$$\int_{0}^{2} (2x - x^2)x \, dx = \int_{0}^{2} (2x - x^2) dx;$$

$$\int_{0}^{\pi/2} [\sin^2 \theta - (1 - \cos \theta)^2] d\theta = \int_{0}^{\pi/2} [\sin \theta - (1 - \cos \theta)] d\theta.$$

Note that the last equation is easily verified without actually integrating by noting from symmetry that $\int_{0}^{\pi/2} \sin^k \theta \, d\theta = \int_{0}^{\pi/2} \cos^k \theta \, d\theta$ for $k = 1, 2$.

Contributed by James C. Kirby of Tarleton State University, Stephenville, TX.

11. Cauchy's negative definite integral

If the integrand is nonnegative, then the definite integral should also be nonnegative. However, the following example (from A.L. Cauchy, *Mémoire sur les intégrales définies,* Seconde partie, III, Exemple II, p. 405–406; Oeuvres (1) 1 (1882) 319–506) led Cauchy to reflect more deeply on the evaluation of the definite integral and to develop his calculus of residues.

Problem. Evaluate

$$\int_{0}^{3\pi/4} \frac{\sin x}{1 + \cos^2 x} dx.$$

Solution. Since the derivative of $\arctan(\sec x)$ is the integrand, the integral is equal to $\arctan(-\sqrt{2}) - \arctan 1 = -\arctan \sqrt{2} - \pi/4$, a negative quantity. ♠

However, it is easy to see that the integrand is defined and nonnegative on the closed interval $[0, 3\pi/4]$. What goes wrong? The function $\arctan(\sec x)$ certainly is an anti-derivative of the integrand, but it has a jump discontinuity at $\pi/2$, and this invalidates an application of the fundamental theorem of calculus. The solution can be patched by using a continuous anti-derivative such as $-\arctan(\cos x)$ or the function $f(x)$ defined by

$$f(x) = \begin{cases} \arctan(\sec x) & 0 \le x < \tfrac{1}{2}\pi \\ \tfrac{1}{2}\pi & x = \tfrac{1}{2}\pi \\ \pi + \arctan(\sec x) & \tfrac{1}{2}\pi < x \le 3\pi/4. \end{cases}$$

Alternatively, one can split the domain of integration into two subintervals $[0, \tfrac{1}{2}\pi]$ and $[\tfrac{1}{2}\pi, 3\pi/4]$ and add the integrals over each. Either way leads to the correct result $(3\pi/4) - \arctan\sqrt{2}$.

I2. A positive vanishing integral

Problem.   Evaluate

$$\int_{-1}^{1} (1 + x^2)^{-1} dx.$$

Solution.   Since the derivative of

$$\frac{1}{2} \arccos \frac{1 - x^2}{1 + x^2}$$

is $(1 + x^2)^{-1}$, the given integral is equal to $\tfrac{1}{2}\arccos 0 - \tfrac{1}{2}\arccos 0 = 0$. ♠

Since both $(1+x^2)^{-1}$ and $\tfrac{1}{2}\arccos(1-x^2)(1+x^2)^{-1}$ are even functions, we should treat the assertion that the first is the derivative of the second with some skepticism. The derivative of the arccos function contains the expression $(4x^2)^{\frac{1}{2}}$, which equals $2x$ when $x > 0$, but equals $-2x$ when $x < 0$. Hence $\tfrac{1}{2}\arccos(1 - x^2)(1 + x^2)^{-1}$ is the antiderivative of $(1 + x^2)^{-1}$ only when $x > 0$; when $x < 0$, the antiderivative is the negative of the arccos function.

Contributed by M. Bencze of Brasov, Romania.

I3. Blowing up the integrand

Suppose that one makes a change of variable $x = u(t)$ in an improper integral $\int_a^\infty f(x)dx$ to obtain $\int_b^\infty g(t)dt$. If, say, $\lim_{x \to \infty} f(x) = 0$, must the function g enjoy the same property? The answer is "no", and this suggests a "blow-up-the-integral" test for showing the divergence of integrals whose integrands vanish at infinity. For example, it is not immediately clear, when $\alpha > 1$, whether the integral

$$\int_2^\infty \frac{1}{(\log x)^\alpha} dx$$

converges. This becomes immediately resolved by making a change of variable: $t = \log x$. The integral becomes

$$\int_{\log 2}^\infty \frac{e^t}{t^\alpha} dt.$$

Since the integrand obviously approaches infinity along with t, the integral diverges. Ronald J. Fischer of Evergreen Valley College in San José, CA, who drew attention to this phenomenon, remarks that it seems surprising that

> an integral whose integrand approaches zero can be replaced by one whose integrand approaches infinity. What seems to be happening is that the substitution $x = e^t$ is forcing large values of x to be replaced by much smaller values of t, so small in fact that the integrand has to blow up in order to retain the divergence of the integral. A sort of exponential scaling is taking place.

> Other examples can be given. The same substitution $x = e^t$ leads to

$$\int_1^\infty \frac{1}{x^\alpha} dx = \int_0^\infty e^{(1-\alpha)t} dt$$

in which an easy analysis can be based on the right side of the equation. When $\alpha = 1$, the integral with integrand $1/x$ is replaced by one with a constant integand. Using instead the substitution $x = e^{e^t}$ leads to

$$\int_2^\infty \frac{1}{x} dx = \int_{\log(\log 2)}^\infty e^t \, dt$$

with an exploding integrand on the right.

14. Average chord length

Problem. Let $\mathcal{E}$ be the closed region in $\mathbf{R}^2$ bounded by the curves of equations $y = x^3$ and $y = \sqrt{x}$. What is the average length of all horizontal chords AB of the region $\mathcal{E}$?

Solution I. The chords in question have endpoints $(x, \sqrt{x})$ and $(x^{1/6}, \sqrt{x})$ for $0 \le x \le 1$, and so the lengths are $x^{1/6} - x$. By the formula for the average value of a function on an interval, we have that the average chord length is

$$\int_0^1 (x^{1/6} - x) dx = \frac{5}{14}.$$

Solution II. The chords have endpoints (x^6, x^3) and (x, x^3) for $0 \le x \le 1$ and so the average chord length is

$$\int_0^1 (x - x^6)dx = \frac{5}{14}.$$

Solution III. The chords have endpoints (y^2, y) and $(y^{1/3}, y)$ for $0 \le y \le 1$, so the average chord length is

$$\int_0^1 (y^{1/3} - y^2)dy = \frac{5}{12}.$$

Shall we go with the majority on this one?

Contributed by Bernard C. Anderson of Manchester, MI.

15. Area of an ellipse

The area enclosed by an ellipse with parametric equations $x = 4\cos\theta$, $y = 3\sin\theta$ is 12π. However, one student determined the area in the following way.

Problem.   Determine the area enclosed by the ellipse with equations

$$x = 4\cos\theta \quad y = 3\sin\theta \quad (0 \le \theta < \pi).$$

Solution.   Using the formula $\int r^2(\theta)d\theta/2$ for area in polar coordinates, one finds the answer to be

$$\int_0^{2\pi} (4^2\cos^2\theta + 3^2\sin^2\theta)d\theta/2 = 25\pi/2. \spadesuit$$

What is wrong? While the polar formula is valid when θ is the polar angle ($\arctan y/x$) for a point on the curve, in the parametrization given, θ is not the polar angle. For example, if $\theta = (1/4)\pi$, then $x = 2\sqrt{2}$ and $y = (3/2)\sqrt{2}$. The polar angle is $\arctan(3/4)$, which is not $(1/4)\pi$.

16. Infinite area but a finite volume

Problem E1151 in the *American Mathematical Monthly* 62 (1955) 121, 581–582, asks for a resolution of the paradox: The area between the curve $y = 1/x$ and the x-axis to the right of the line $x = 1$ is infinite. Yet the volume generated by rotating this area about the x-axis is π. Thus, it would require an infinite amount of paint to cover the area; yet the volume, which completely contains and surrounds the area, can be filled with π cubic units of paint! Was this the inspiration behind the film *Infinite acres* (Melvin Henriksen; Modern

Learning Aids, Rochester, 1965; 11 min.) which treats amusingly a solid of revolution with finite volume and infinite surface area?

Leonard Gillman, of Austin, TX, comments:

In the summer of 1964 at Stanford, Mel [Henriksen] and I were in the same general group, I at SMSG and he at the film project. I am quite certain that yes he definitely got his inspiration for the film *Infinite Acres* from the famous solid of finite volume but infinite surface area (now often known as Gabriel's Horn). I believe Mel heard of Gabriel's Horn from me, as we were close friends and colleagues at Purdue from 1952 to 1958, where we discussed all topics relevant to mathematics professors.

I also believe I heard the example the day it was invented. This was on a Monday in Spring 1941 at Columbia University, where I had just begun graduate studies. The mathematics faculty lunched together at the faculty club every Monday, and on this particular day, P.A. Smith recounted that in his 11 o'clock calculus class, he had presented the example and then on the spur of the moment, had added. "So you see, you can fill it with water but you can't paint it." Clearly, he thought that was pretty clever, and so did his luncheon colleagues, including J.F. Ritt, B.O. Koopman, E.R. Lorch, Walter Strodt, and I think Howard Levi. Just as they all burst out laughing, Strodt murmured, "Why not make it out of glass and fill it with paint?" Levi told me the same story the same afternoon as we were walking home.

By the way, I have wondered ever since why no one ever mentions the simpler example one dimension down. Forget the curve $y = 1/x^2$ and calculus. Just make a pasture consisting of the rectangles of height $1/2^n$ on the interval $[n-1, n]$. To show that the area is finite, you deal with the simplest convergent geometric series of all time, and to see that the perimeter is infinite, just listen to the statement of the problem: no computing of any sort is needed.

I7. An Euler equation

Problem.   Solve the equation

$$x^2 y'' - 5xy' + 9y = 0.$$

Solution.   Try the solution $y = e^{rx}$ to get $[(rx)^2 - 5(rx) + 9]e^{rx} = 0$. Thus $rx = (5 \pm i\sqrt{11})/2$. Hence the complex-valued function

$$y = e^{5/2 \pm i\sqrt{11}/2} = e^{5/2}\left(\cos\frac{\sqrt{11}}{2} \pm i\sin\frac{\sqrt{11}}{2}\right)$$

is a solution. Since the differential equation is linear, the real and imaginary parts of this complex solution are also solutions. Thus, the equation has two constant solutions:

$$y = e^{5/2}\cos\frac{\sqrt{11}}{2} \quad \text{and} \quad y = e^{5/2}\sin\frac{\sqrt{11}}{2}.$$

But if $y = c$ is a solution, then $y' = y'' = 0$, so $x^2 y'' - 5xy + 9y = 9c$, which is zero only if $c = 0$. It follows then that $\sqrt{11}/2$ must be equal to $\pi/2$ for the first solution or π for the second. ♡

Based on a student solution, contributed by Bart Braden of Northern Kentucky University in Highland Heights.

18. Solving a second-order differential equation

Upsilon.   Here is a method for getting a particular integral for the differential equation

$$y'' - 4y' + 3y = (4x - 2)e^x.$$

Suppose that the solution is of the form $y(x) = u(x)e^x$ and write the equation in the form

$$(D^2 - 4D + 3I)(ue^x) = (4x - 2)e^x$$

where D is the differential operator and I is the identity. Using the result that, for any polynomial $p(D)$ in D and any number k, $p(D)(ue^{kx}) = e^{kx} \cdot p(D + kI)(u)$, we have that

$$(4x - 2)e^x = (D - 3I)(D - I)(ue^x) = e^x \cdot (D - 2I)Du.$$

Cancelling the factor e^x yields

$$4x - 2 = (D - 2I)Du . \tag{1}$$

Differentiate to get

$$4 = (D - 2I)D^2 u \tag{2}$$

$$0 = (D - 2I)D^3 u. \tag{3}$$

We make $D^3u = 0$. From equation (2) we find that $D^2u = -2$, and from (1) that $Du = -2x$. This is all we need, so we can take $u = -x^2$ for the particular integral $y = -x^2 e^x$. ♣

Omicron. I can see how any particular integral can be put in the form $y = ue^x$, since we can take $u = ye^{-x}$. But you just pull the condition $D^3u = 0$ out of thin air. This is not a necessary consequence of equation (3).

U. Will you agree that $y = -x^2 e^x$ is a particular integral?

O. (Checking) Sure.

U. Well then, the proof of the pudding is in the eating. I try on the condition $D^3u = 0$ and it gives me a solution. What more is needed?

O. Maybe you were lucky. You have no right to suppose that $D^3u = 0$ even if it does work out in practice. You need advance assurance that such a u exists.

U. Suppose, more generally, I need a particular integral of $p(D)y = q(x)e^{kx}$ where p and q are polynomials. Trying $y = ue^{kx}$ leads to

$$p(D + kI)u = q . \tag{4}$$

Solving this for u is equivalent to solving the original equation. Now, just keep differentiating this until the right side vanishes, getting $p(D + kI)D^m u = 0$ for m equal to the sum of the number of differentiations and the multiplicity of k as a root of p. I can set $D^m u = 0$ and then work my way back up through a succession of equations to (4), getting lower order derivatives of u in turn. Isn't this enough of a prospectus for an argument?

O. You're sure that the values of the derivatives are consistent?

19. Power series thinning

Problem. Prove that

$$\frac{1}{2}e^x = \sum_{n=0}^{\infty} \frac{x^n}{n!} - \left(1 + x + \frac{x^4}{4!} + \frac{x^5}{5!} + \frac{x^8}{8!} + \frac{x^9}{9!} + \cdots \right) .$$

Solution. Consider the differential equation $y'' + y = e^x$. Solving this in the usual way to obtain a complementary function and particular integral yields

$y(x) = a \cos x + b \sin x + (1/2)e^x$. However, solving in series yields

$$y(x) = a\left(1 - \frac{x^2}{2} + \frac{x^4}{4!} + \cdots\right) + b\left(x - \frac{x^3}{3!} + \frac{x^5}{5!} - \frac{x^7}{7!} + \cdots\right)$$
$$+ \left(\frac{x^2}{2!} + \frac{x^3}{3!} + \frac{x^6}{6!} + \frac{x^7}{7!} + \cdots\right)$$
$$= a \cos x + b \sin x + \left(\frac{x^2}{2!} + \frac{x^3}{3!} + \frac{x^6}{6!} + \frac{x^7}{7!} + \cdots\right).$$

Subtraction of $a \cos x + b \sin x$ from both solutions yields the result. ♠

Of course, the constants a and b are not the same in both solutions. In the series solution, $a = y(0)$ and $b = y'(0)$. The general solution with these initial conditions is actually

$$y(x) = \left(a - \frac{1}{2}\right) \cos x + \left(b - \frac{1}{2}\right) \sin x + \frac{e^x}{2}.$$

Contributed by David Rose of Oral Roberts University in Tulsa, OK.

Chapter **8**

CALCULUS: MULTIVARIATE AND APPLICATIONS

I. Variable results with partial differentiation

Let $F(x, y) = (x + y)^2$. Set $x = u - v$ and $y = u + v$. Then

$$\frac{\partial x}{\partial v} = -1, \frac{\partial y}{\partial v} = 1, \frac{\partial F}{\partial x} = \frac{\partial F}{\partial y} = 2(x + y).$$

By the chain rule,

$$\frac{\partial F}{\partial v} = \frac{\partial F}{\partial x}\frac{\partial x}{\partial v} + \frac{\partial F}{\partial y}\frac{\partial y}{\partial v} = -2(x + y) + 2(x + y) = 0.$$

But, from the definition, $F(u, v) = (u+v)^2$ whence $\partial F/\partial v = 2(u+v) = 2y$. We seem to have found that $y = 0$. ♣

In the two computations of $\partial F/\partial v$, the meaning of the variables u and v is not consistent. We can clarify the first by writing the function of u and v as the composite of two functions:

$$g(u, v) = (u - v, u + v)$$

and

$$F(x, y) = (x + y)^2$$

so that $(F \circ g)(u, v) = (2u)^2$. What we are really computing is the partial derivative of $F \circ g(u, v) = 4u^2$ with u and v as independent variables:

$$\frac{\partial}{\partial v}(F \circ g)(u, v) = \frac{\partial}{\partial v}(4u^2) = 0.$$

As for the second computation, it is really a statement about the partial of F with respect to its second variable and is equivalent to $\partial(x + y)^2/\partial y = 2(x + y)$.

Contributed by Hugh Thurston of the University of British Columbia in Vancouver.

2. Polar paradox?

Anyone who has taken multivariate calculus has seen the polar conversion formulae

$$r^2 = x^2 + y^2 \quad \tan\theta = \frac{y}{x} \quad x = r\cos\theta \quad y = r\sin\theta.$$

Let us assume that all quantities are defined (i.e., no division by zero). On the one hand, since $r = \sqrt{x^2 + y^2}$ we have

$$\frac{\partial r}{\partial x} = \frac{x}{\sqrt{x^2 + y^2}} = \frac{x}{r} = \frac{r\cos\theta}{r} = \cos\theta.$$

On the other hand, though, $r = x\sec\theta$, so that $\partial r/\partial x = \sec\theta$. What is going on here? ♣

In multivariate differential calculus, the evaluation of partial derivatives depends on the variables that are considered to be independent. (Readers who have taken a course in thermodynamics will appreciate this issue.) We can accept $\partial r/\partial x = \sec\theta$ on the understanding that x and θ are the independent variables upon which r depends.

Normally however, we consider (x, y) as a pair of independent variables upon which r and θ depend. Accordingly, the dependence of $r = x\sec\theta$ on x would be mediated through $\sec\theta$ as well as x. We would then have

$$\frac{\partial r}{\partial x} = \sec\theta + x\sec\theta\tan\theta\frac{\partial\theta}{\partial x} = \sec\theta\left(1 + y\frac{\partial\theta}{\partial x}\right).$$

Since $\tan\theta = y/x$, it follows that $(\sec^2\theta)(\partial\theta/\partial x) = -y/x^2$, so that $\partial\theta/\partial x = -y/r^2$. Thus

$$\frac{\partial r}{\partial x} = \frac{\sec\theta}{r^2}(r^2 - y^2) = \frac{x^2\sec\theta}{r^2} = \frac{(x^2\sec^2\theta)}{r^2}\cos\theta = \cos\theta,$$

in agreement with the first computation.

Contributed by Therese Shelton of Southwestern University, Georgetown, TX.

3. Polar increment of area

Since $x = r\cos\theta$ and $y = r\sin\theta$, it follows that $dx = dr\cos\theta - r\sin\theta d\theta$ and $dy = dr\sin\theta + r\cos\theta d\theta$. Taking the product yields

$$dxdy = \cos\theta\sin\theta(dr)^2 + r\cos^2\theta dr d\theta - r\sin^2\theta dr d\theta - r^2\cos\theta\sin\theta(d\theta)^2.$$

Considering $(dr)^2$ and $(d\theta)^2$ negligible, we find that

$$dA = dx\,dy = r(\cos^2\theta - \sin^2\theta)dr\,d\theta. \quad \clubsuit$$

Putting aside the question of neglecting $(dr)^2$ and $(d\theta)^2$, the expression $dx\,dy$ can be regarded as the area of an infinitesimal rectangle with sides parallel to the coordinate axes determined by increments in the variables x and y. The standard element of area in polar coordinates, namely $r\,dr\,d\theta$, is the area of an infinitesimal region determined by increments in the coordinates r and θ; it is not the rectangle just described but rather the region bounded by rays with separation $d\theta$ and circular arcs with radial difference dr.

Contributed by Peter Jarvis and Paul Shuette of Georgia College in Milledgeville.

4. Evaluating double integrals

Consider the following computation:

$$\int_0^6 \int_0^{\sqrt{5-x}} 4y\,dy\,dx = \int_0^6 (10 - 2x)\,dx = 60 - 36 = 24.$$

To check, reverse the order of integration. Note that increasing the upper limit on the outside integral sign decreases the value of the integral! Here are some additional examples in which the flimflam is a little more concealed:

$$\int_0^2 \int_0^{\sqrt{3-x^2}} 6y\,dy\,dx = \int_0^2 (9 - 3x^2)\,dx = 10$$

$$\int_0^{\pi/2} \int_0^{\sqrt{2-3\cos\theta}} 2r\,dr\,d\theta = \int_0^{\pi/2} (2 - 3\cos\theta)\,d\theta = \pi - 3. \quad \clubsuit$$

Contributed by Leonard Gillman of Austin. TX.

5. One-step double integration

Here is a working of a double integral by a student in a Calculus 3 class. The answer is correct.

$$\int_0^2 \int_0^2 (16 - x^2 - 2y^2)\,dx\,dy = \left| 16xy - \frac{x^3}{3}y - 2x\frac{y^3}{3} \right|_{x=y=0}^{x=y=2}$$

$$= 64 - \frac{16}{3} - \frac{32}{3} = 48.$$

In general,

$$\int_a^b \int_c^d x^r (y^s \, dy) \, dx = \left[\frac{b^{r+1} - a^{r+1}}{r+1} \cdot \frac{d^{s+1} - c^{s+1}}{s+1} \right]$$

for $r, s \in \mathbf{N}$. When $a = c = 0$, the student's strategy of integrating both x^r and y^s simultaneously will give the correct answer.

Contributed by James C. Kirby of Tarleton State University, Stephenville, TX.

6. The converse to Euler's theorem on homogeneous functions

Recall that a real-valued function f of several variables is *homogeneous of degree* p if for all vectors $\mathbf{x} = (x, y)$ and scalars t with $\mathbf{x}$ and $t\mathbf{x}$ in the domain of f, $f(t\mathbf{x}) = t^p f(\mathbf{x})$. Euler's theorem provides that, if f is differentiable and homogeneous of degree p on a domain S, then for all $\mathbf{x} \in S$, $\mathbf{x} \cdot \nabla f(\mathbf{x}) = pf(\mathbf{x})$.

The converse is often established by this argument. For fixed $\mathbf{x}$, differentiate $g(t) = f(t\mathbf{x}) - t^p f(\mathbf{x})$, and use the assumption $\mathbf{x} \cdot \nabla f(\mathbf{x}) = pf(\mathbf{x})$ to show that g satisfies $tg'(t) - pg(t) = 0$. For $t \neq 0$, the general solution is $g(t) = C|t|^p$ for some constant C. Since $g(1) = 0$, $C = 0$ and so $f(t\mathbf{x}) = t^p f(\mathbf{x})$ whenever $t \neq 0$. Also, it follows directly from the hypothesis that $f(0) = 0$. Thus, f is homogeneous of degree p. ♣

However, consider the following example. Let $S = \mathbf{R}^2$, and define

$$f(x, y) = \begin{cases} xy^2, & y \geq 0 \\ y^3, & y < 0. \end{cases}$$

Since f has continuous partial derivatives on S, it is differentiable on S. Furthermore $\mathbf{x} \cdot \nabla f(\mathbf{x}) = 3f(\mathbf{x})$ on S. However, letting $t = -1$ and $\mathbf{x} = (0, 1)$ reveals that f is not homogeneous on S.

A second example shows that a problem can arise, even if t is restricted to being nonnegative. Let S be the plane with the upper half of the unit circle removed. Define

$$f(x, y) = \begin{cases} xy^2, & y \geq 0 \text{ and } x^2 + y^2 > 1 \\ y^3, & \text{elsewhere.} \end{cases}$$

Then f is differentiable and $\mathbf{x} \cdot \nabla f(\mathbf{x}) = 3f(\mathbf{x})$ but f is not homogeneous on S.

A correct statement of the converse can be found in Exercise 6 on page 186 of A. Taylor & W. Mann, *Advanced calculus* (Wiley, 1972). Formulated for two variables, the result is: suppose that $f(x, y)$ is defined and differentiable in an open region S, and suppose that

$$x\frac{\partial f}{\partial x} + y\frac{\partial f}{\partial y} = pf$$

at each point of the region. Then, for $(x, y) \in S$, the relation $f(tx, ty) = t^p f(x, y)$ holds in any interval $t_0 < t < t_1$ provided $t_0 \geq 0$, $t = 1$ is in the interval, and for all t, $(tx, ty) \in S$.

To prove this, fix (x, y) and define $g(t) = f(tx, ty)$. Use the hypothesis to prove that $tg'(t) = pg(t)$, and then to infer that $g(t)t^{-p}$ is constant (i.e., depends only on x and y).

Contributed by Robert Cacioppo of Northeast Missouri State University in Kirksville.

7. The wrong logarithm

A student was asked to find the proportion of carbon 14 remaining in a sample after 9000 years. The rate of decay, $k = \ln 2/5730 = 0.000121$, had already been derived, and the student found the factor $e^{-(0.000121)(9000)}$ by punching the calculator keys

$$[0.000121][+/-][\times][9000][=][\text{Shift}][\text{Log}]$$

to get 0.08, far too small. Advised that she should have used [Shift] [Ln], the student protested that she had used the [Log] key in all the other problems and got correct answers.

The problems were of the type: given a growth or decay function $P(t) = P_0 e^{kt}$ along with the values $P(0) = P_0$ and $P(t_1) = P_1$, determine the constant k and evaluate $P_2 = P(t_2)$. The professor pointed out to the student: "Actually, you are not *really* doing anything wrong. The formula $P(t) = P_0 e^{kt}$ is also $P(t) = 10^{Kt}$ where $10^K = e^k$. And so, since you were not given k ahead of time, your procedure *assumed* the 10-form, found K, and everything after that was perfectly correct." The student responded, "In that case, why do we have to use that weird number e at all? Why can't we just stick with 10?"

Contributed by Eric Chandler of Randolph-Macon Woman's College in Lynchburg, VA.

8. The conservation of energy according to Escher

Theorem.   There is a function $z = f(x, y)$ defined for $(x, y) \neq (0, 0)$ such that

$$\nabla f = \frac{-y}{x^2 + y^2}\mathbf{i} + \frac{x}{x^2 + y^2}\mathbf{j}.$$

Proof.

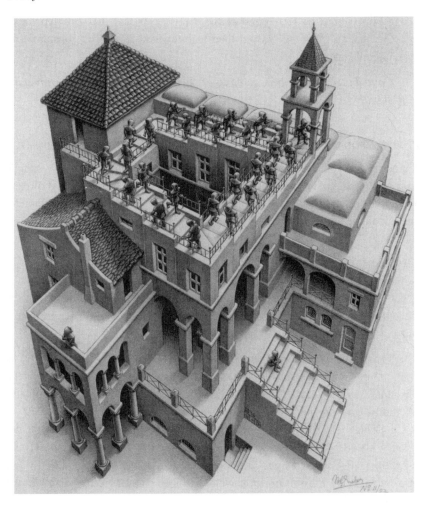

FIGURE 8.8

Ascending and Descending by M. C. Escher.

9. Calculating the average speed

Bill Simpson of Michigan State University in East Lansing offers a glimpse into the mind of a beginning calculus student. He is assigned the problem of finding the average speed for the time period measured from $t = 0$ to $t = 3$ seconds for a particle whose displacement in feet is given by $P(t) = 3t^2 + 4t$. The student thinks:

> Now let's see; he's asking for the speed—that's always a derivative in calculus, so we calculate $v(t) = P'(t) = 6t + 4$. Now what? Well—I'm given $t = 0$ and $t = 3$, so these must be used somehow—probably to evaluate the speed at the start and finish of the trip. Let's try that: $v(0) = 4$ and $v(3) = 22$. But this gives me two results and clearly the problem wants only one. Oh, I see now! He wants the "average" velocity—so clearly the answer is $\frac{1}{2}(4 + 22) = 13$ feet per second.

The answer is correct. Will this always be so? Specifically, what are the functions $f(x)$ for which

$$\frac{f(b) - f(a)}{b - a} = \frac{f'(b) + f'(a)}{2}$$

for any choice of values a and b? This is a nice question for a class in elementary differential equations.

10. Maximizing a subtended angle

The following problem, originally distributed to an educational discussion group by Richard Askey, appeared on the Lower Michigan Mathematics Competition for 1998. The solution is essentially from one of the candidates.

Problem. Let $O = (0,0)$ and $Q = (1,0)$. Find the point P on the line $y = x + 1$ for which the angle OPQ is a maximum.

Solution. Let $P(x, x + 1)$ be a typical point on the line. For the required point, we must have $0 \le x \le 1$. In the triangle POQ, $\angle OPQ$ is maximized when $\angle POQ + \angle PQO$ is minimized. Since the sine function is monotonic, we have to minimize $\sin \angle POQ + \sin \angle PQO$. By the Law of Sines, this is equivalent to minimizing

$$|OP| + |PQ| = \sqrt{x^2 + (x + 1)^2} + \sqrt{(x - 1)^2 + (x + 1)^2}$$
$$= \sqrt{1 + 2x + 2x^2} + \sqrt{2 + 2x^2}.$$

This happens when $x = 0$, so P must be the point $(0, 1)$. ♡

11. Hanging oneself with a minimum of rope

The following problem appears in a widely used calculus text; the solution is from the solutions manual.

Problem. Two vertical poles PQ and ST are secured by a rope PRS going from the top P of the first pole to a point R on the (level) ground between the poles and then to the top S of the second pole. Show that the shortest length of such a rope occurs when $\angle PRQ = \angle SRT$.

Solution. Let $d = |QT|$ and $x = |QR|$. We can minimize $f(x) = |PR| + |RS| = x \sec\theta_1 + (d-x)\sec\theta_2$, where $\theta_1 = \angle PRQ$ and $\theta_2 = \angle SRT$. Then $f'(x) = \sec\theta_1 - \sec\theta_2 = 0 \Leftrightarrow \theta_1 = \theta_2$, since $0 < \theta_1 \le \theta_2 < (\pi/2)$. So the shortest rope occurs when $\theta_1 = \theta_2$. ♠

But, of course, θ_1 and θ_2 also vary with x. It is interesting to note that the heights of the poles do not enter into the solution. The quickest valid solution to the problem avoids calculus by minimizing the path from P to S', where T is the midpoint of SS', and uses the reflection principle. However, if we wish to keep x as the independent variable, a solution can be found as follows. Let $a = |PQ|$, $b = |ST|$, $u = |PR|$, $v = |RS|$, so that we need to minimize $f(x) = u + v$. Since $a^2 + x^2 = u^2$ and $b^2 + (d-x)^2 = v^2$, we have that $x = uu'$ and $-(d-x) = vv'$, whence

$$f'(x) = \frac{x}{u} - \frac{d-x}{v} = \cos\theta_1 - \cos\theta_2.$$

As x increases from 0 to d, θ_1 decreases from $\pi/2$ and θ_2 increases to $\pi/2$. Thus $f'(x)$ increases from a negative quantity to a positive quantity, and vanishes when $\theta_1 = \theta_2$. Thus the length of the rope is minimized when $\theta_1 = \theta_2$.

Contributed by Don Kemp of the South Dakota State University in Brookings.

12. Throwing another fallacy out the window

Here is another problem from a standard calculus text with a solution from the solutions manual.

Problem. A child standing 20 feet from the base of a silo attempts to throw a ball into an opening 40 feet from the point of release [the diagram indicates 40 feet up from the base]. Find the minimum initial speed and the corresponding angle at which the ball must be thrown to go into the opening.

Solution. $r(t) = t(v_0 \cos\theta)\mathbf{i} + (tv_0 \sin\theta - 16t^2)\mathbf{j}$. [At the opening,] $dy/dt = v_0 \sin\theta - 32t = 0$, when $t = (v_0 \sin\theta)/32$. At that time,

$$x(t) = v_0 \cos\theta \left(\frac{v_0 \sin\theta}{32} \right) = \frac{v_0^2}{64} \sin 2\theta = 20$$

and

$$y(t) = \frac{v_0^2 \sin^2\theta}{32} - 16 \left(\frac{v_0^2 \sin^2\theta}{32^2} \right) = \frac{v_0^2 \sin^2\theta}{64} = 40.$$

Solving these equations, we have that $\tan\theta = 4$ or $\theta \sim 76°$ and $v_0 = 4\sqrt{170}$ ft/sec. ♠

Assume that the ball starts at $(0,0)$ and that the opening is located at (d, h). The path of the ball is given parametrically by

$$(x(t), y(t)) = \left(u(\cos\theta)t, u(\sin\theta)t - \frac{g}{2}t^2 \right)$$

where g is the acceleration due to gravity, θ is the angle of throw, and u is the initial speed of the ball. The ball reaches its maximum height at time $t_c = u\sin\theta/g$; if this occurs at the point (d, h), then $\tan\theta = 2h/d$, so $u = \sqrt{g(4h^2 + d^2)/2h}$. This represents the alleged minimum speed. When h is small relative to d, this is a large quantity.

A correct approach to the problem involves considering u as a function of θ. Since the path of the ball is given by the equation

$$y = f(x) = -\frac{gx^2}{2u^2 \cos^2\theta} + (\tan\theta)x,$$

we can set $(x, y) = (d, h)$ and solve for u^2:

$$u^2(2\cos\theta)(d\sin\theta - h\cos\theta) = gd^2.$$

The critical angle satisfies the relation $\tan 2\theta = -d/h$ or $\tan\theta = (h + \sqrt{d^2 + h^2})/d$. If h is small, then θ is close to $45°$ as one would expect. If d is small, then the correct and incorrect angles will be close to $90°$.

We can check the answer by looking at extreme cases. If h/d is near 0, intuition reveals the error: the text's θ is small and its u is huge. If a baseball player makes a casual, leisurely throw or a basketball star puts up a gentle jump shot, are they trying to have their targets at the vertices of the balls' flight? Certainly not. If we really desire a low-speed trajectory, the vertex will precede the target point, and by a significant distance if the target is not "high and near". This is a nice example to encourage students to look back on their answers and check them against extreme cases and intuition.

Contributed by Paul Deiermann and Rick Mabry of Louisiana State University at Shreveport, with some of the analysis provided by David Abrahamson of Rhode Island College in Providence.

13. Generalizing an approach to the radius of curvature

Paul Deiermann and Rick Mabry of the Louisiana State University in Shreveport fly the following trial balloon about treating radius of curvature. In most texts nowadays, the curvature of a space curve at a point is defined using the length of the limiting ratio of the rate of change of the unit tangent vector to the corresponding arc length traversed by the curve. The curvature of a circle is then shown to be the reciprocal of the circle's radius, so it becomes reasonable to define the radius of curvature as the reciprocal of the curvature.

In two dimensions, one can take the reverse approach by deriving the radius of curvature without any mention of curvature. The idea is to look at the intersection points (assuming they exist) of normal lines to the curve at nearby points on the curve, then let one point on the curve approach a fixed point on the curve and see where the limiting intersection points converge. The limiting point will be the center of the osculating circle, and the distance from this point to the fixed point will be the radius of curvature.

To carry the above idea over to space curves, one may not have an intersection point of the normal lines to our curve, so just pick the "closest thing" to an intersection point, i.e., the points at which the normal lines are nearest to each other. Other than this, the above procedure should work. Right?

14. The lopsided uniform rod

A fairly routine application of the integral in standard calculus texts is that of computing the center of mass of a lamina. Consider a lamina in the shape of a sector of a circle of angle θ strictly between 0 and $\pi/2$ and or radius R. Assuming that the lamina has uniform unit density and lies in the first quadrant with the lower edge along the x-axis, we find that its center of mass is at

$$\left(\frac{2R\sin\theta}{3\theta}, \left(\tan\frac{\theta}{2} \right) \left[\frac{2R\sin\theta}{3\theta} \right] \right).$$

Let θ tend to 0. Physically, it is "clear" that the limiting object is a straight rod of uniform density 1. Mathematically, it is true that the limits of the coordinates of the center of mass are $(2R/3, 0)$. Thus, the center of mass of a uniform rod of length R is two thirds of the way along its length. ♣

This argument from Mark G. Leeney of the University College of Cork in Ireland drew the following response from Carlton A. Lane of Hillsborough Community College in Tampa, FL. He is skeptical about relying on physical intuition for valid predictions about non-physical things such as the objects in this situation. What happens as the sector angle approaches $\pi/2$? Then the centroid approaches the correct position for a quarter circle. Now replace the sector by a right triangular lamina with vertices at $(0,0)$, $(R,0)$, and $(R, R\tan\theta)$. The abscissa of the centroid is $(2/3)R$, independently of θ. As $\theta \to 0$, we again get the lopsided rod, while $\theta \to \pi/2$ yields a lopsided slab $\{(x,y) : 0 \le x \le R, y \ge 0\}$. While the mathematical error is manifest, the erroneous physical assumption is masked because the limiting figure is in essence a vertical filament for which the difference between $(2/3)R$ and $(1/2)R$ is negligible in the physical sense.

However, in this situation, we could even review our physical intuition. When $\theta \neq 0$, for the sector, the amount of mass in an infinitesimal arc of radius r is proportional to r, while for the triangle the amount of mass in an infinitesimal strip of length x is proportional to x. In either case, as $x \to 0$, it is reasonable to posit that the relative distribution of mass persists to yield a density of the rod that is proportional to x, say kx. Then the mass of the rod would be $\frac{1}{2}kR^2$ and its total moment from the end O, $\frac{1}{3}kR^3$.

Chapter 9

LINEAR AND MODERN ALGEBRA

I. A proof that O = I

Choose a positive integer n and let S be the set of all real solutions $(x_1, x_2, \ldots, x_n, x_{n+1})$ of the equation

$$x_1 + x_2 + \cdots + x_n + x_{n+1} = 1. \tag{$*$}$$

Since any choice of $x_1, x_2, \ldots, x_n$ determines exactly one value of x_{n+1} for which ($*$) holds, the mapping $f : \mathbf{R}^{n+1} \to \mathbf{R}^n$ defined by $f(x_1, x_2, \ldots, x_{n+1}) = (x_1, x_2, \ldots, x_n)$ maps S one-to-one onto $\mathbf{R}^n$. Moreoever, f preserves addition and scalar multiplication (defined as usual). It follows that S is isomorphic to $\mathbf{R}^n$ and is hence a vector space. In particular, S must contain the zero vector; that is, the vector $(0, 0, \ldots, 0, 0)$ satisfies ($*$) so $0 = 1$.

Contributed by David Bloom, Brooklyn College, New York.

2. Matrices and the TI-8I graphics calculator

Several years ago, Constance J. Gardner of Cuyahoga Community College of Parma, OH was teaching precalculus, incorporating the use of the TI-81 graphics calculator. While preparing a lesson, she inadvertently created an example that can be incorrectly interpreted when using this calculator. Here it is.

Solve the following system of equations:

$$x + 2y + 3z = 20$$
$$2x + 3y + 4z = 29$$
$$3x + 4y + 5z = 38.$$

The system has an infinite number of solutions, the ordered triples of the form $(c-2, -2c+11, c)$.

Her calculator, purchased in the fall of 1990, evaluated $[A]^{-1}[B]$ as $[-4\ 0\ -4]^T$, which disconcertingly is not even a correct particular answer.

What is the problem? The calculator gave det $[A]$ as $-3E-12$, even though the determinant is really zero. Since the calculator gave a nonzero value to the determinant, it came up with an inverse and thus an answer. She tried the problem on a second TI-81 graphics calculator purchased in February, 1992. This time, the calculator evaluated $[A]^{-1}[B]$ as $[3\ 10\ 4]$, another wrong answer. The determinant was computed to be $-4E-12$.

Gardner comments that we still need to urge our students to understand the basic ideas behind any area of mathematics. Without a quick check of either of these "solutions," we might never have realized they were incorrect. She suggests that if students are solving equations by the $[X] = [A]^{-1}[B]$ method, they should have the calculator evaluate the determinant of A. If the determinant is nearly zero, then an exact computation of it should be made. If it indeed vanishes, then the TI-81 should be used to help perform a Gaussian elimination to solve the system.

3. The Schwarz-Cauchy Inequality

One linear algebra student used the associativity of the inner product to reverse a standard inequality. Let u and v be two vectors of a real inner product space; since the inequality we derive is trivial when $v = 0$, we suppose that $v \neq 0$ and that $\lambda = u \cdot v / v \cdot v$. Then

$$0 \le (u - \lambda v) \cdot (u - \lambda v) = u \cdot u - 2u \cdot \lambda v + \lambda^2 v \cdot v$$
$$= u \cdot u - \frac{2u \cdot (u \cdot v) \cdot v}{v \cdot v} + \frac{(u \cdot v)^2 (v \cdot v)}{(v \cdot v)^2}$$
$$= u \cdot u - \frac{2(u \cdot u) \cdot (v \cdot v)}{v \cdot v} + \frac{(u \cdot v)^2}{v \cdot v} = -u \cdot u + \frac{(u \cdot v)^2}{v \cdot v}$$

whence $(u \cdot u)(v \cdot v) \le (u \cdot v)^2$. ♡

Assuming associativity of the inner product actually leads to equality:

$$(u \cdot u) \cdot (v \cdot v) = u \cdot (u \cdot v) \cdot v = u \cdot (v \cdot u) \cdot v = (u \cdot v) \cdot (u \cdot v).$$

Contributed by M. J. de la Puente of Universidad Complutense in Madrid, Spain.

4. An entrance examination question

A letter having to do with an issue of justice in *The Mathematical Intelligencer* 19:3 (Summer, 1997, page 4) contains the following problem, posed on a college entrance examination in South Korea in 1995:

> Three nonzero vectors A, B and C in three-dimensional Euclidean space satisfy the following inequality
>
> $$\|xA + yB + zC\| \geq \|xA\| + \|yB\|$$
>
> for all real numbers x, y and z. Show that the three vectors are perpendicular to each other.

While the conclusion can certainly be established from the hypothesis, it turns out that there are no three nonzero vectors satisfying the hypothesis.

5. Nonsquare invertible matrices

Here is a student argument about the invertibility of nonsquare matrices.

Proposition. All nonsquare matrices are invertible.

Proof. (1) If A is not square, then det A does not exist.

(2) If det A does not exist, then it certainly cannot equal zero.

(3) If det A is not zero, then A is invertible. ♠

Which step of the proof is in error? Step (1) and Step (2) pass muster. So the difficulty resides in Step (3). But here the student was quoting directly from the text. So maybe the text writer was not sufficiently careful in formulating the result. Instead of the statement, *If* det A *is not zero, then* A *is invertible*, the text should have had *If* det A *is nonzero, then* A *is invertible*. In the later case, the hypothesis makes it clear that the determinant has a value and it is not zero.

This contribution from Allen J. Schwenk of Western Michigan University in Kalamazoo drew a response from Joan A. Carr of the Department of Technical Mathematics, National Technical Institute for the Deaf, Rochester Institute of Technology:

> Allen J. Schwenk has pointed out that the language of mathematics texts is sometimes imprecise. The recounting of his student's false proof that nonsquare matrices must be invertible demonstrates the misunderstanding that can take place as a result. I agree that the verbal statement "det A is not zero" is at fault. However, I must

take exception to the assertion that the symbolic statement "det $A \neq 0$" is equally bad. Since $\neq$ is in this context a relation on $\mathbf{R} \times \mathbf{R}$, det $A \neq 0$ is true if and only if det A is *real* and nonzero, and hence only if matrix A is square. Even though a more explicit hypothesis might benefit the student, the following statement of the theorem is technically correct: if det $A \neq 0$, then A is invertible.

A comment from Peter Renz of the Massachusetts Institute of Technology:

> A good joke always teaches us something, and this business about a nonsquare matrix having an inverse because its (nonexistent) determinant is not equal to 0 is a case in point. The positive version of this theorem is more general, clearer, and follows more directly from the proof. A matrix over a commutative ring R is invertible if and only if its determinant is invertible in R. A ring theorist would surely state the result this way. See for example page 96 of *Basic Algebra I*, Second edition, by Nathan Jacobson, published by W.H. Freeman and Company, New York, 1985. The corollary for fields with det $A \neq 0$ follows on the same page.

6. An inversion conundrum

Let n be an integer exceeding 1 and let $AX = B$ represent a system of n linear equations in n unknowns, where A is an $n \times n$ square matrix and X and B are column n-vectors. This has the solution

$$X = A^{-1}B.$$

Let α be the nonzero scalar given by $\alpha = X^T B$, where X^T represents the transpose of X, and let M be the matrix

$$M = \alpha^{-1}XX^T.$$

Then

$$MB = \alpha^{-1}XX^TB = X[\alpha^{-1}X^TB] = X.$$

Thus $X = MB$ as well as $X = A^{-1}B$. Therefore $M = A^{-1}$. However, M is a matrix of rank 1 and hence is singular. This contradicts the fact that it is the inverse of the matrix A. ♣

The argument works except for the assertion that $M = A^{-1}$. All we can say is that $(M - A^{-1})B = O$, at which point we are confounded by the possibility of zero divisors. As an example take

$$A = \begin{pmatrix} 2 & 1 \\ 1 & 1 \end{pmatrix} \qquad \text{and} \qquad B = \begin{pmatrix} 3 \\ 2 \end{pmatrix}$$

so that

$$A^{-1} = \begin{pmatrix} 1 & -1 \\ -1 & 2 \end{pmatrix}, \qquad X = \begin{pmatrix} 1 \\ 1 \end{pmatrix}, \qquad \alpha = 5, \qquad M = \frac{1}{5} \begin{pmatrix} 1 & 1 \\ 1 & 1 \end{pmatrix}.$$

Then

$$M - A^{-1} = \frac{1}{5} \begin{pmatrix} -4 & 6 \\ 6 & -9 \end{pmatrix}$$

and this indeed annihilates B without itself vanishing.

Contributed by Barry D. Ganapol of the University of Arizona in Tucson.

7. The Cayley-Hamilton Theorem

Theorem. Let $p(\lambda) = \det(A - \lambda I)$ be the characteristic polynomial of the square matrix A. Then $p(A) = 0$.

Proof. $p(A)$ is found by replacing each occurrence of λ in the polynomial $p(\lambda)$ with A. Doing this yields $p(A) = \det(A - AI) = \det O = 0$, as desired. ♠

The proof cannot be correct, since the last equation starts off with a matrix and ends up with a scalar. W. Watkins of the California State University at Northridge tells his students the following:

> You've picked the wrong time to substitute A for λ. But if you insist on making the substitution before taking the determinant, it has to be done more carefully. Let's look at the 2×2 case
>
> $$A = \begin{pmatrix} a & b \\ c & d \end{pmatrix}.$$
>
> If we substitute A for λ in the characteristic matrix
>
> $$A - \lambda I = \begin{pmatrix} a - \lambda & b \\ c & d - \lambda \end{pmatrix},$$
>
> we get the block matrix
>
> $$\begin{pmatrix} aI - A & bI \\ cI & dI - A \end{pmatrix}.$$

Now take the block determinant to get

$$\text{DET}\begin{pmatrix} aI - A & bI \\ cI & dI - A \end{pmatrix} = (aI - A)(dI - A) - bcI$$

$$= A^2 - (a+d)A + (ad - bc)I.$$

The Cayley-Hamilton theorem states that this *matrix* is zero.

Siu Man-Keung of the University of Hong Kong uses the "proof" as a pedagogical strategy in his linear algebra class. He writes:

> First I show the class a letter of Cayley to Sylvester [given on pages 213–214 in the article "Cayley's anticipation of a generalized Cayley-Hamilton theorem" by Tony Crilly in *Historia mathematica* 5 (1978) 211–219]. It has the nice feature of, through a concrete 2×2 case, illustrating the content of the theorm to a beginner. I then repeat the words of Cayley in 1858: "The determinant, having for its matrix a given matrix less than the same matrix *considered as a single quantity* involving the matrix unity, is equal to zero." (italics mine). With this preparation, I produce the "proof" and (while some students are nodding approvingly!) ask the class whether it is correct. (If it was, the theorem would be a triviality!) After that (sometimes in the next lecture) I explain what is wrong, placing emphasis on the properties of polynomials. Then I give a sort of "brute-force" proof, and finally (for the better students) explain that the "joke-proof" can be turned into a rigorous proof by regarding $A - XI$ as a polynomial over the $n \times n$ matrices. In the course of this proof the "brute-force" part would slip in once more. For the two times I treated the theorem this way, the class liked it and it stimulated much discussion among the better students.

In the *American Journal of Mathematics* 41 (1919) 266–278, H. B. Phillips proved a generalization of the Cayley-Hamilton theorem: Let $A_1, \ldots, A_k$ and $B_1, \ldots, B_k$ be $n \times n$ matrices where the B_i are pairwise commutative and $A_1 B_1 + \cdots + A_k B_k = O$. Define a polynomial in k indeterminates by $p(x_1, \ldots, x_k) = \det(A_1 x_1 + \cdots + A_k x_k)$. Then $p(B_1, \ldots, B_k) = O$, the zero matrix.

8. All groups are simple

Theorem. *All groups are simple.*

Proof. For a group G with identity e, we define the product of any two subsets of G by the formula $AB = \{ab : a \in A, b \in B\}$. Clearly the subset

$\{e\}$ is an identity for this operation of "subset multiplication." On the other hand, if H is a normal subgroup of G, then it is well known that G/H (the set of cosets of H in G) is a group under the *same* operation and that H is the identity thereof. Since a set cannot have more than one identity with respect to a given operation, it follows that H must equal $\{e\}$. Thus, G has no nontrivial normal subgroup. ♡

    Contributed by David M. Bloom, of Brooklyn College of the City University of New York.

    The following two items are from an article "Errors and misconceptions in college level theorem proving" by Annie Selden of the Tennessee Technological University in Cookeville, and John Selden, published in *Proceedings of the Second International Seminar, Misconceptions and Educational Strategies in Science and Mathematics*, vol. III, Cornell Univesity, 1987, pages 457–470.

9. Groups with separate identities

Theorem. *Let G_1 and G_2 be two groups contained in a semigroup S such that $G_1 \cap G_2$ is nonempty. If e_1 is the identity of G_1 and e_2 is the identity of G_2, then $e_1 = e_1 e_2 e_1$.*

Proof. There is an element g in $G_1 \cap G_2$. Then $e_1 g = g = e_2 g$. Since g is a group element, g^{-1} exists, and $e_1 g g^{-1} = e_2 g g^{-1}$, so that $e_1 e_1 = e_2 e_2$ or $e_1 = e_2$. Multiplying on the right and left by e_1 yields $e_1 = e_1 e_1 e_1 = e_1 e_2 e_1$. ♠

    The inverses of g in G_1 and G_2 may not be the same. Here is a correction of the argument. Let g_1 be the inverse of g with respect to the identity e_1 in G_1. Then

$$e_1 g = e_2 g \implies e_1 = e_1 g g_1 = e_2 g g_1 = e_2 e_1$$

so $e_1 = e_1^2 = e_1 e_2 e_1$.

10. The least common multiple order

Theorem. *If a commutative group has an element of order 2 and an element of order 3, then it must have an element of order 6.*

Proof. Let g and h be elements of orders 3 and 2 respectively. Then $g^3 = h^2 = e$. Since $h^6 g^6 = (h^2)^3 (g^3)^2 = e$, the subgroup generated by hg must be

$\{hg, h^2g^2, h^3g^3, h^4g^4, h^5g^5, h^6g^6\}$ which simplifies to $\{hg, g^2, h, g, hg^2, e\}$. So hg has order 6. ♠

At issue here is just an oversight on the part of the prover; it needs to be verified that the six elements are distinct.

II. The number of conjugates of a group element

Proposition. Let G be a finite group. Define the relation R on G by: gRh iff there is an x in G such that $g = x^{-1}hx$. Then R is an equivalence relation, and the number of elements in the equivalence class containing g is the index in G of G_g, the subgroup of all x in G for which $g = x^{-1}gx$.

Proof. With g fixed, on the set $S_g = \{x^{-1}gx : x \in G\}$ of conjugates of g, define an operation $*$ by

$$(x^{-1}gx) * (y^{-1}gy) = (xy)^{-1}g(xy).$$

This operation is clearly associative, $e^{-1}ge$ is an identity and xgx^{-1} is an inverse of $x^{-1}gx$. Hence S_g is a group under $*$. Define the mapping fromkk G onto S_g by $x \to x^{-1}gx$; this clearly is onto and homomorphic. The kernel of this homomorphism is $G_g = \{y \in G : y^{-1}gy = g\}$. By the fundamental isomorphism theorem, G/G_g is isomorphic to the group S_g, from which we conclude that $|S_g| = |G|/|G_g|$. ♠

To be the kernel of a homomorphism, G_g must be normal, which we know is not always the case. So we should look more closely at the definition of the homomorphism. Is it well defined? A conjugate $u = x^{-1}gx$ of g could be implemented by different elements x. Suppose that

$$u = x_1^{-1}gx_1 = x_2^{-1}gx_2 \qquad \text{and} \qquad v = y_1^{-1}gy_1 = y_2^{-1}gy_2.$$

Can we be sure that $(x_1y_1)^{-1}g(x_1y_1) = (x_2y_2)^{-1}g(x_2y_2)$ so that $u * v$ is uniquely defined?

For example, let G be the symmetric group on three elements (with multiplication of permutations from left to right) and $g = (12)$. Then $S_g = \{e, (12)\}$ and

$$(13)^{-1}(12)(13) = (123)^{-1}(12)(123) = (23)$$

and

$$(23)^{-1}(12)(23) = (132)^{-1}(12)(132) = (13).$$

But

$$[(123)(132)]^{-1}(12)[(123)(132)] = e(12)e = 12$$

while

$$[(13)(23)]^{-1}(12)[(13)(23)] = (123)^{-1}(12)(123) = (23)$$

so that $(23) * (13)$ is not well-defined.

   Contributed by David Mead of the University of California at Davis.

12. Even and odd permutations

David Mead of the University of California, Davis came across a published note purporting to prove that a permutation cannot be expressed as both an even number and an odd number of transpositions. This led to a fairly extended interchange in the *College Mathematics Journal* that unearthed a subtle point about the representation of permutations. First the argument:

   Assuming that both an even and odd representation is possible leads to a representation

$$e = (a_1 b_1)(a_2 b_2) \cdots (a_k b_k)$$

of the identity as a product of oddly many transpositions, with $1 \le a_r < b_r \le n$. When $a_r \ne 1$, write $(a_r b_r) = (1 a_r)(1 b_r)(1 a_r)$ to get $e = (1 c_1)(1 c_2) \cdots (1 c_h)$ with h odd and $c_i \ne 1$ for all i. Since the identity permutation maps each symbol s $(1 < s \le n)$ onto itself, the latter expression for e must have each $(1s)$ occurring evenly often, which contradicts the oddness of h. ♣

   However, $e = (12)(13)(12)(13)(12)(13)$ is an example in which (12) and (13) each occur an odd number of times.

   David Berman of the University of New Orleans in Louisiana then wrote in to say that he had used a similar strategy to establish the result and pointed to his proof that the identity cannot be written as an odd product of transpositions in *Mathematical Gazette* 62 (1978) 211–212. This employed an induction argument to show that, if the identity was represented as a product of *switches*, i.e., permutations $(i, i+1)$ of adjacent ordinals, then the number of switches had to be even. The induction step was based on the observation that the switch $(n-1, n)$ had to appear an even number of times in order for n to be restored to its original position at the end of the line.

   George Mackiw of Loyola College in Baltimore, MD weighed in with the complaint that Berman's proof "suffers from the same flaw" as the earlier proof, and asked us to consider $e = (12)(23)(12)(23)(12)(23)$ in which the number of switches involving the number 3 is odd. While at first blush Mackiw's criticism of Berman's proof seems to be valid, a closer analysis vindicates Berman. There is a subtle twist involved, so let us examine it in more detail.

In Berman's proof, the elements being permuted are $\{a_1, a_2, \ldots, a_n\}$. A *switch* is described as a transposition of the form (a_p, a_{p+1}) and it is noted that any transposition (a_p, a_{p+k}) can be expressed as the product of $2k - 1$ switches. An induction argument can be marshaled to prove that the identity is not expressible as an odd number of switches. This is clear if $n = 2$ and we suppose it for $n \geq 2$.

> Consider the identity permutation of $\{a_1, a_2, \ldots, a_{n+1}\}$ expressible as the product of switches. Since each element ends in its original position, the number of switches involving a_{n+1} is even. Now consider the switches not involving a_{n+1}. A switch involving a_{n+1} has no effect on the order (relative to each other) of the other n elements. Therefore, the switches not involving a_{n+1} constitute the identity permutation on $\{a_1, a_1, \ldots, a_n\}$. So by the induction hypothesis, there is an even number of such switches. Thus, the total number of such switches is even.

How can we reconcile this with Mackiw's purported counterexample? A permutation is a one-to-one function defined on the set $\{1, 2, 3, \ldots\}$, and products of permutations (read from left to right) are compositions of these functions. Thus, (12) represents the function $1 \to 2$, $2 \to 1$, $k \to k$ $(k \geq 3)$; (23) the mapping $1 \to 1$, $2 \to 3$, $3 \to 2$, $k \to k$ $(k \geq 4)$, so that the composite $(12)(23)$ is the mapping $1 \to 3$, $2 \to 1$, $3 \to 2$, $k \to k$ $(k \geq 4)$. Note that in this usage, the domain $\{1, 2, 3, \ldots\}$ is an unordered set.

However, Berman's argument depends on lining up in order the elements to be permuted, and this order is changed with each successive permutation. *Switches* are transpositions of two numbers that happen to be adjacent when the switch takes place. Thus, if we start with the ordered set $[1, 2, 3]$, the first permutation (12) yields $[2, 1, 3]$ and is indeed a switch. But, (23) applied to the order $[2, 1, 3]$ yields $[3, 1, 2]$ and is not a switch as 2 and 3 are not adjacent.

Let us define $\langle ab \rangle$ as that permutation that interchanges the numbers (whatever they may be) that happen to be in the ath and bth positions from the left. Consider the product $\langle 12 \rangle \langle 23 \rangle \langle 12 \rangle \langle 23 \rangle \langle 12 \rangle \langle 23 \rangle$. If we start with the order $[1, 2, 3]$, these permutations are all switches and produce successively the ordered sets $[2, 1, 3]$, $[2, 3, 1]$, $[3, 2, 1]$, $[3, 1, 2]$, $[1, 3, 2]$, $[1, 2, 3]$. The composite is indeed the identity, but the reader will observe that four switches involve the number 3, without affecting the order of 1 and 2, and two switches interchange the numbers 1 and 2, while leaving 3 immobile.

Mackiw's counterexample fails, since it does not involve switches in Berman's sense, and if we reinterpret his example to refer to transposition of position rather than element, it does not challenge Berman's argument. This is an issue that could easily escape attention. For example, users of John B. Fraleigh's textbook, *A First Course in Abstract Algebra* (5th ed., Addison-Wesley, 1994) might take note of this discussion when assigning Problem 28 on page 112, which is based on Berman's argument.

Daniel J. Bernstein of Emeryville, CA, draws attention to an alternative definition of the parity of a permutation. Define σ to be the number of pairs (i, j) whose order is reversed under the permutation σ, i.e., $i < j$ and $i\sigma > j\sigma$. Then inv $(\sigma\tau) \equiv$ inv σ + inv τ (mod 2) and inv $((a, a+1, \ldots, b-1, b)) = b-a$ for any cycle of consecutive integers. Since $(a, a + 1, \ldots, b - 1, b)(a, b) = (a, a + 1, \ldots, b - 1)$, we conclude that inv τ is odd for any transposition τ and that the permutation σ is even in the usual sense if and only if inv σ is even.

Mackiw himself uses the determinant to show that the parity of the number of transpositions in a product representation of a permutation is invariant. One associates to each permutation its permutation matrix; the result follows from the facts that a transposition matrix has determinant -1 and that taking determinants respects products. There is a danger of circularity in this approach, but this can be obviated by defining determinant without recourse to permutations. He comments that, in his opinion, "the major advantage of this 'linear algebraic' proof of invariance of parity is that it makes sense to students—they accept it."

How large is the set of degenerate real symmetric matrices?

Peter D. Lax, Courant Institute of Mathematical Sciences, NY, NY

Consider the space of all $n \times n$ real, symmetric matrices. Such a matrix is degenerate if and only if it has a multiple eigenvalue.

The eigenvalues of a matrix are the roots of its characteristic polynomial. The condition for a polynomial to have multiple roots is that its discriminant vanishes. This discriminant is a polynomial in the coefficients of the characteristic polynomial, which in turn are polynomials in the entries of the matrix. Thus, the condition for degeneracy is a *single* equation for the entries of the matrix; this shows that the degenerate matrices form a manifold (algebraic variety) of *codimension one* in the space of all real, symmetric matrices.

This conclusion however is false; it it were true, then an arbitrarily chosen line $A + tB$, with A and B real symmetric matrices chosen at random, would intersect the manifold of degenerate matrices. But this can be disproved by a simple computer experiment, using *MatLab Maple*; pick A and B and plot the eigenvalues of $A + tB$ as functions of t on a sufficiently dense set of points t. You will observe that as t varies, the eigenvalues of $A + tB$ approach each other, but in the last minute turn aside and fail to intersect. This phenomenon is known as "avoidance of crossing." Its explanation was given as long ago as 1927 by von Neumann and Wigner, who showed that the degenerate matrices form a subvariety of *codimension two*. Here is their argument:

> Every $n \times n$ real symmetric matrix has n real eigenvalues and n real eigenvectors that can be chosen to be orthonormal. Use the normalized eigenvectors and eigenvalues to parametrize the space of real, symmetric matrices; how many parameters are there? The first normalized eigenvector has $n - 1$ parameters; the second, constrained to be orthogonal to the first one, has $n - 2$ parameters, and so on; altogether we have $(n - 1) + (n - 2) + \cdots + 1 = \frac{1}{2}n(n - 1)$ parameters. To this we have to add the n parameters represented by the n eigenvalues, getting altogether $\frac{1}{2}n(n - 1) + n = \frac{1}{2}n(n + 1)$ parameters; this equals the dimension of the space of symmetric $n \times n$ matrices.
>
> Take now the set of degenerate symmetric matrices; say the last two eigenvalues are equal. Since there is no distinguished eigenvector in two-dimensional eigenspace corresponding to the degenerate eigenvalue, we lose one of the parameters. We lose a second parameter because there are only $n - 1$ distinct eigenvalues; this shows that the degenerate symmetric matrices can be described by two fewer parameters than the space of all symmetric matrices.

The resolution of the paradox is an interesting problem in real algebraic geometry for the reader. (Hint: try the case $n = 2$.) There is an illustration of the avoidance of crossing on the dust-jacket of Lax's text, *Linear algebra*, published by Wiley (1997).

Other items.   See also Item 13 (Why Wiles' proof of the Fermat theorem is wrong) in Chapter 1.

Chapter 10

ADVANCED UNDERGRADUATE
MATHEMATICS

I. Troublemakers

Two separate footnotes appear on page 139 of the 1925 textbook *A Practical Treatise on Fourier's Theorem and Harmonic Analysis for Physicists and Engineers* by Albert Eagle (Longman's, Green & Co., London), where the author treats the proposition that a Fourier series converges to the average of the right and left limits of its parent function:

> A case in point is the series $\sin 1/t + \frac{1}{2}\sin 2/t + \frac{1}{3}\sin 3/t + \cdots$. As t increases towards $+\infty$, this series, as we have seen, approaches the limit of $+\frac{\pi}{2}$; while when t is negative and moves towards $-\infty$ the series approaches the limit of $-\frac{\pi}{2}$ while actually *at* $t = \pm\infty$ the series is indeterminate and can assume any value from -1.8519 to $+1.8519$. Such behaviour, however, is not difficult to detect, and in the case of infinite series, is always bound up with the fact that *at* such a point the series does not begin to converge till after an infinite number of terms.

The second footnote follows on the derivation of the nth partial sum for the Fourier series of $f(x)$ at t as

$$\frac{1}{\pi}\int_{-n'\pi}^{n'\pi} f\left(t + \frac{v}{n'}\right) \frac{\sin v}{2n' \sin \frac{v}{2n'}}\, dv$$

where $n' = n + \frac{1}{2}$. Eagle points out that when $n' = \infty$, this becomes

$$\frac{1}{\pi}\int_{-\infty}^{+\infty} f(t)\frac{\sin v}{v}\, dv = f(t)$$

provided that $f(t + \frac{v}{n'})$ is sensibly constant for all values of v below a reasonably large quantity, say V, when n' is very large. This is generally expressed by saying that $f(t)$ must not have an infinite number of maxima and minima or discontinuities in a finite space in the neighbourhood of the value t considered; for if so, $f(t + \frac{v}{n'})$ could fluctuate in magnitude during a finite change in v when n' was infinite: and so we could tell nothing about the value of the integral

$$\int f\left(t + \frac{v}{n'}\right) dv.$$

This is the reason why so many books dealing with Fourier's series continually repeat the condition that the function must not have an infinite number of maxima and minima. We have generally omitted specifying this condition, since no practical function ever does behave in such a manner. Such behaviour is exclusively confined to functions invented by mathematicians for the sake of causing trouble. ◇

2. The countability of the reals

Theorem.   *The set of all real numbers is countable.*

Proof.   Consider the set **R** of reals as an index set. For each $r \in \mathbf{R}$, let S_r be a convergent sequence of rationals whose limit is r. The sets S_r are nonvoid; the intersection of any two of these sets is at most finite (so that they are "almost disjoint"); their union is countable, being a subset of the rationals. Hence, the index set must also be countable. ♡

All that we have here is an example to illustrate the perhaps counter-intuitive possibility that a union of uncountably many sets whose pairwise intersections are at most finite can actually be only countable.

At about the same time as this item appeared in the *College Mathematics Journal* (November, 1989), the 1989 Putnam competition was being written. Problem B4 posed the following question: Can a countably infinite set have an uncountable collection of nonempty subsets such that the intersection of any two of them is finite?

3. The plane constitutes an uncountable set

Theorem.   *The real plane is not the union of countably many lines.*

Proof. Suppose, if possible, that the plane is the union of the lines L_k ($k = 1, 2, \ldots$). For each point P in the plane, associate a positive integer, namely, the smallest positive integer i for which $P \in L_i$. Now rotate all the lines $90°$ about some point D, sending L_k to L'_k. Use the new lines to associate to P a second positive integer, the smallest j for which $P \in L'_j$. We obtain a function mapping P to (i, j), with P the unique point of intersection of L_i and L'_j. Thus, there is an injection of the plane into a countable set $N \times N$, so that the plane is countable. But this is false, and so the stated result must hold. ♠

Indeed, the result of the theorem is true, but the proof has a hole. Suppose that P corresponds to (i, j), so that it lies on L_i and the image L'_j of a line L_j after a $90°$ rotation. It may happen that L_i and L'_j coincide and that some other point Q corresponds to the same pair of indices.

Contributed by John Wilker of the University of Toronto in Ontario.

4. A consequence of the nearness of rationals to reals

Proposition. $1 = 0$.

Proof. We accept the fact that the set of rationals in $[0, 1]$ is both countable and dense in the interval. Suppose $\epsilon > 0$. Let $\{r_1, r_2, r_3, \ldots\}$ be an enumeration of the rationals in $[0, 1]$. For each positive integer i, let $J_i = \{x : 0 \leq x \leq 1, r_i - 2^{i+1}\epsilon < x < r_i + 2^{i+1}\epsilon\}$. Since the rationals are dense in $[0, 1]$, each element of the interval belongs to at least one of the intervals J_i. Thus $[0, 1] \subseteq \cup_{i=1}^{\infty} J_i$, so that $1 = \text{length}[0, 1] \leq \sum_{i=1}^{\infty} \text{length } J_i = \sum_{i=1}^{\infty} 2^{-i}\epsilon = \epsilon$. Since this holds for each $\epsilon > 0$, the result follows. ♠

The covering certainly does seem to leave no number in $[0, 1]$ unblanketed. However, an example can be given of a number specifically excluded from such a covering. Suppose that $\{r_n\}$ is the following listing of rationals:

$$\left\{ 0, 1, \frac{1}{2}, \frac{1}{3}, \frac{2}{3}, \frac{1}{4}, \frac{3}{4}, \frac{1}{5}, \frac{2}{5}, \frac{3}{5}, \frac{4}{5}, \frac{1}{6}, \frac{5}{6}, \cdots \right\},$$

where each fraction is listed according to the size of its denominator and subordinately its numerator when written in lowest terms. For a given positive ϵ, the nth rational r_n is included in an open interval of length $\epsilon/2^n$ centered on $\{r_n\}$. If ϵ is less that $1/5$, then it turns out that no interval includes $\sqrt{2}/2$. Recall the following theorem from Section 6.3. of *Introduction to the Theory of Numbers* (3rd ed., Wiley, 1972) by I. Niven and H.S. Zuckerman.

Theorem. *Let h/k be a rational in $(0,1)$ written in lowest terms. Then $\sqrt{2}/2$ is not included in the interval*

$$\left(\frac{h}{k} - \frac{1}{4k^2}, \frac{h}{k} + \frac{1}{4k^2}\right).$$

Proof.   The proof by contradiction is straightforward. If, for some h, k,

$$\frac{h}{k} - \frac{1}{4k^2} < \frac{\sqrt{2}}{2} < \frac{h}{k} + \frac{1}{4k^2},$$

then squaring each term, multiplying by $2k^2$, and subtracting $2h^2$ yields

$$-\frac{h}{k} + \frac{1}{8k^2} < k^2 - 2h^2 < \frac{h}{k} + \frac{1}{8k^2}.$$

The left side is greater than $-h/k > -1$ while the right side is less than $h/k + 1/k \le 1$, so that $k^2 - 2h^2 = 0$. But this contradicts the irrationality of $\sqrt{2}$. ♡

Note that, for $\epsilon \le 1/5$,

$$\frac{\epsilon}{2(2^n)} \le \frac{1}{4(n+1)^2}.$$

Since the denominator k of the nth rational $r_n = h/k$ does not exceed $n+1$, it follows that the interval with center r_n and length $\epsilon/2^n$ fits inside the interval

$$\left(\frac{h}{k} - \frac{1}{4k^2}, \frac{h}{k} + \frac{1}{4k^2}\right)$$

and so does not contain $\sqrt{2}/2$.

Fallacy devised by Mark Lynch of Millsaps College in Jackson, MS. Example of uncovered number provided by W. Christopher Lang of Indiana University Southeast in New Albany.

5. A universal property of real subsets

Theorem. *Every subset of the reals is a G_δ (i.e., a countable intersection of open sets).*

Proof.   Denote the set by A. For each positive integer n and each x in A, let $I(x, n)$ be the open interval $\{u : |x - u| < 1/n\}$. Then $J(n) = \cup\{I(x, n) : x \in A\}$ is an open set containing A. Since $\{x\} = \cap_n I(x, n)$, it follows that $A = \cap_n J(n)$ so that A is a G_δ. ♠

To test the conclusion that A is the intersection of the $J(n)$, look at the situation in which $A = \mathbf{Q}$ and each $J(n) = \mathbf{R}$.

An immediate consequence of item 5 is that every real subset is Lebesgue measurable. This is corroborated by the following result, but then read on further!

6. A topological spoof

We require a background result:

Lemma. *Any uncountable G_δ subset of **R** contains a nowhere dense closed set C of measure zero that can be mapped continuously onto $[0, 1]$.*

This is Lemma 5.1 on page 23 of J.C. Oxtoby, *Measure and category* (2nd edition, Springer, 1980). The construction in the proof makes it clear that C is homeomorphic to the classical Cantor set K contained in $[0, 1]$.

Theorem. *The set of real numbers is countable.*

Proof. It suffices to show that the set I of irrationals is countable. Suppose I is uncountable. Since I is the intersection of complements of rational singletons $\{x\}$, I is a G_δ. By Oxtoby's proof, I contains a subset C which, being homeomorphic to K, is compact, and hence closed and bounded. The complement U of C is a disjoint union $(-\infty, a_0) \cup (b_0, \infty) \cup \cup_{k=1}^\infty (a_k, b_k)$ which contains the rationals. Since Q is dense in **R**, there can be no gaps between consecutive interval components of U. Thus, $C = \cup_{k=0}^\infty \{a_k, b_k\}$ is countable, and with it, the Cantor set K. But K and **R** have the same cardinality, so **R**, hence I, is countable. This contradicts the hypothesis and so establishes the result. $\heartsuit$

Contributed by David A. Rose of East Central University in Ada, OK. Cf. item 9 in this chapter.

7. Is there a nonmeasurable set?

Theorem. *Every subset of the closed interval $[a, b]$ is Lebesgue measurable.*

Proof. Let a subset of $[a, b]$ be given and suppose that g is its characteristic function (taking the value 1 on the set and 0 elsewhere). It suffices to show that g is integrable. Let F be the space of all functions on $[a, b]$ with values in $[0, 1]$ endowed with the topology of pointwise convergence. F is compact. We can find in F a countable dense sequence S consisting of all simple functions, defined with respect to partitions of $[a, b]$ by rational points, which assume rational values. Choose a subsequence of S which converges to g. Since every member of S is Lebesgue integrable, so also is g by the dominated convergence theorem. ♠

The topology on F is not metrizable, so while any sequence has a convergent sub*net*, the subnet need not be a subsequence. For the necessary theory, consult John L. Kelley, *General Topology* (Van Nostrand, Princeton, NJ, 1955; reprinted by Springer-Verlag, New York), page 136, or Sze-Tsen Hu, *Elements of General Topology* (Holden-Day, San Francisco, 1964), page 75.

8. Is there a nonmeasurable set?

Theorem. *There exists a subset of* $[0, 1]$ *which is not Lebesgue measurable.*

Proof. Assume that every set $X \subseteq [0, 1]$ has a measure $m(X)$. Clearly, $m(X)$ is a number in $[0, 1]$ that may or may not be in X. Define $B = \{m(X) : m(X) \notin X\}$. Then B is a subset of $[0, 1]$. We have that $m(B) \in B$ if and only if $m(B) \notin B$, which is ridiculous. Hence our assumption cannot be sustained. ♠

The difficulty is that the mapping $X \longrightarrow m(X)$ is many-one. Consider for example the number

$$\frac{1}{3} = m\left(\left[\frac{1}{6}, \frac{1}{2}\right]\right) = m\left(\left[\frac{1}{2}, \frac{5}{6}\right]\right).$$

Thus $m(B)$ could indeed belong to B since $m(B)$ could equal $m(C)$ for some set C which does not contain it.

Contributed by A.R. Freedman of Simon Fraser University in Burnaby, BC.

9. Is there a function continuous only on the rationals?

Many elementary analysis courses present an example of a real function defined on the reals which is continuous exactly at the irrational points. Remarkably, it is not possible to find a function which is continuous exactly on the rationals. While this can be handled using the heavy machinery of the Baire category theorem, there is an elementary proof of this fact due to Volterra (*Giornale di matematiche* 19 (1881) 76–87 = *Opera matematiche*, Vol. 1, 7–15). More accessible perhaps is Problem 41 on page 322 of *Real Variables* by J.M.H. Olmsted which asks for an argument that the discontinuities of a real-valued function of a real variable form an F_σ set. Those who do not have these at hand can ponder the following.

Theorem. *There is no real-valued function defined on the reals, the set of whose points of continuity consists exactly of the rationals.*

Proof. Suppose that f is continuous at each rational. Define

$$\omega(x) = \inf_{\delta > 0} \sup\{f(u) - f(v) : |x - u| < \delta, \ |x - v| < \delta\}.$$

Then $\omega(x) = 0$ for rational x and $U_n = \{x : \omega(x) < 1/n\}$ is open for each positive integer n. Thus U_n is the union of countably many open intervals and this union contains every rational. If (a, b) and (c, d) are two consecutive intervals in U_n with $b \le c$, then $b = c$ (otherwise, there would be a rational in (b, c)). Hence, the complement of U_n consists only of the endpoints of the intervals, a countable set. Therefore $\{x : \omega(x) = 0\} = \cap_{n=1}^\infty U_n$ has a countable complement and hence is uncountable. But f being continuous whenever $\omega(x)$ vanishes, has uncountably many points of continuity and so is continuous at points other than the rationals. The result follows. ♠

Even though U_n is the union of countably many open intervals, these intervals could be arbitrarily small in length. Similar to the rationals in the reals, they could be positioned in such a way that some intervals do not have a nearest neighbor.

10. The continuum hypothesis

Professor E.P.B. Umbuggio showed the existence of a cardinal between the cardinalities of the integers and the reals as follows: Obviously $1 < \aleph_0$. Therefore, $\aleph_0 < \aleph_0^{\aleph_0}$. Moreover, $\aleph_0 < 2^{\aleph_0}$. Therefore, $\aleph_0^{\aleph_0} < (2^{\aleph_0})^{\aleph_0} = 2^{(\aleph_0 \cdot \aleph_0)} = 2^{\aleph_0} = \mathfrak{c}$, whence $\aleph_0 < \aleph_0^{\aleph_0} < \mathfrak{c}$. Consult Problem E1979 in the *American Mathematical Monthly* 74 (1967) 438; 75 (1968) 783. ◊

In the *Monthly*, Eric Rosenthal (at that time in high school) notes that in general for cardinals $a < b$ does not imply $a^e < b^e$ for $e \ne 0$. Otherwise, we get the following contradiction for $2 < \mathfrak{c}$:

$$2^{\aleph_0} < \mathfrak{c}^{\aleph_0} = 2^{(\aleph_0)\aleph_0} = 2^{\aleph_0 \cdot \aleph_0} = 2^{\aleph_0}$$

so $2^{\aleph_0} < 2^{\aleph_0}$.

II. A heavy-duty proof that I = 0

The function $\exp(\exp z)$ is entire (i.e., analytic in the whole complex plane and thus representable by an everywhere convergent power series). By Picard's Little Theorem, there is at most one value which it does not assume. It omits 0 because exp does. It also omits 1, again because exp omits 0. Therefore 1 and 0 must be the same number. ♣

For a reference to the cited theorem, consult Walter Rudin, *Real and Complex Analysis*, 3rd edition, McGraw Hill, New York, 1987, p. 331. Note that the function $\exp(\exp z)$ *does* assume the value 1 when $\exp z = 2k\pi i$ for some integer k, i.e., when $z = \log(2k\pi) + \frac{1}{2}\pi i$ in the case $k > 0$ and $z = \log(-2k\pi) - \frac{1}{2}\pi i$ in the case $k < 0$.

Contributed by Richard Parris of Phillips Exeter Academy, New Hampshire.

12. All complex numbers are real

An arbitrary complex number $z = re^{i\theta}$ is not only real, but nonnegative. This is clear when $\theta = 0$. When $\theta \neq 0$, let $\alpha = e^{i\theta}$. Then

$$\alpha^{2\pi/\theta} = (e^{i\theta})^{2\pi/\theta} = e^{2\pi i} = 1,$$

from which it follows that

$$\alpha = (\alpha^{2\pi/\theta})^{\theta/2\pi} = 1^{\theta/2\pi} = 1.$$

The result follows.

Contributed by Walter Reno, while a student at Montana State University in Billings.

13. Opening the floodgates

The following gem is presented by F. Riesz in his monograph, *Les systèmes d'équations linéaires à une infinité d'inconnues* (Gauthier-Villars, 1952), pages 15–18. He then proceeds to analyze the situation.

Proposition. Consider the infinite system of equations in the unknowns $x_1, x_2, \ldots$:

$$a_1^r x_1 + a_2^r x_2 + a_3^r x_3 + \cdots + a_k^r x_k + \cdots = 0 \quad (r = 0, 1, 2, \ldots)$$

where the a_k are complex numbers satisfying $|a_{k+1}| > |a_k|$ and $\lim |a_k| = \infty$. If there is any solution to the system with all x_k nonzero, then every sequence satisfies the system.

Proof. Let $x_k = u_k (k = 1, 2, \ldots)$ satisfy the system. Then $x_k = a_k^m u_k$ (all k) gives a solution for each fixed positive integer m. Therefore, for any entire function $f(z) = c_0 + c_1 z + c_2 z^2 + \cdots$, $x_k = f(a_k) u_k$ determines a solution.

Suppose that $v_1, v_2, \ldots, v_k, \ldots$ is an arbitrary complex sequence. We shall obtain our result by demonstrating the existence of an entire function f for which $f(a_k) = v_k/u_k$ (all k). To do this, use the Weierstrass theorem to find an entire function $F(z)$ whose only zeros are simple ones at the a_k and the Mittag-Leffler theorem to find a meromorphic function $R(z)$ whose only poles are simple ones at the a_k with residues $v_k/(u_k F'(a_k))$. Now set $f(z) = F(z)R(z)$. ♡

For an account of the theory cited, consult Walter Rudin, *Real and complex analysis*, third edition (McGraw-Hill, 1987), pages 273–274, 301–306.

Chapter **II**
PARTING SHOTS

I. Ibn Qurra

Here is a multiple-choice question on the history of mathematics. The dates of birth and death of the Arab mathematician Thabit Ibn Qurra, who translated and commented on Greek higher mathematics, are

(A) 826–901   (B) 833–902   (C) 836–901

(D) 836–911   (E) All of the above.

The correct answer is (E). See Al Abdullah Al-Daffa', *The Muslim contribution to mathematics* (Humanities Press, 1977). The dates (A)–(D) are given, respectively, on pages 44, 13, 59, and 86.

2. Reading a calculator display

Sandra Z. Keith of St. Cloud State University in Minnesota writes:

> A student differentiating $f(x) = \cos^2 x$ obtained the answer $f'(x) = 2\cos x^- \sin x$, which I graded wrong, with my typical lecture on parentheses errors in the margin. When he protested that his answer was the same as mine, I challenged him to plug in numbers on his calculator. Punching the "negative" button on his calculator, he did not need to register a multiplication; the graphics display on his calculator, a TI-85 (or a TI-81), uses this format, the minus sign at a higher level, to denote $2\cos x(-\sin x)$, a language he had incorporated into his written work.

3. Infallibility of a symbolic manipulation program

Dean Clark of the University of Rhode Island in Kingston notes that *Theorist* is a highly and justifiably acclaimed manipulation program that uses a propositional calculus thought to be incapable of producing incorrect conclusions if the initial premises are correct. Square icons signal the correct premises which the user types in. Thereafter, triangular icons (stylizing the three dots of "therefore") signal the correct conclusions obtained using various manipulation commands. For example, the following premises include one identity and two definitions, so they are trivially correct. The rest speaks for itself.

$$\square \ \sqrt{-1} = \sqrt{-1} \qquad \square \ 1 = a \qquad \square \ -1 = b \ .$$

It is absolutely critical to type $1 = a$, $-1 = b$ instead of $a = 1$, $b = -1$. Otherwise, *Theorist* will get wise to our swindle within three steps and issue a "Dialogue Box" with a warning to introduce the imaginary constant i. Now introduce a transformation rule:

$$\boxed{\text{Transform}} \quad \sqrt{\frac{x}{y}} \quad \text{into} \quad \frac{\sqrt{x}}{\sqrt{y}}.$$

Now begins the chain of manipulations starting with the identity $\sqrt{-1} = \sqrt{-1}$. Click on the equal sign, and use *Select In* followed by *Apply* to divide both -1's by 1:

$$\triangle \ \sqrt{\frac{-1}{1}} = \sqrt{\frac{-1}{1}}.$$

Highlight $-1/1$ on the left, and use *Apply* to multiply numerator and denominator by (-1):

$$\triangle \ \sqrt{\frac{(-1)(-1)}{(-1)\cdot 1}} = \sqrt{\frac{-1}{1}}.$$

Highlight $(-1)(-1)/(-1)1$, *Select In*, and use *Simplify* to get

$$\triangle \ \sqrt{\frac{1}{-1}} = \sqrt{\frac{-1}{1}}.$$

Using the *hand cursor*, make the substitutions

$$\triangle \ \sqrt{\frac{a}{b}} = \sqrt{\frac{b}{a}}.$$

Activate the *Transformation Rule* and use the *hand cursor* to cross-multiply:

$$\triangle \ \frac{\sqrt{a}}{\sqrt{b}} = \frac{\sqrt{b}}{\sqrt{a}}$$
$$\triangle \ a = b.$$

Finally, perform two separate *hand cursor* substitutions to obtain

$$\triangle \ 1 = -1.$$

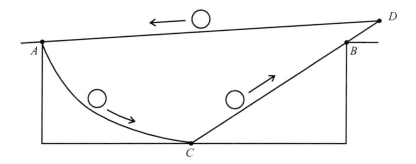

FIGURE **11.4**

4. A perpetual motion machine

In the figure, points A and B are at the same height above point C, which is halfway between them. The ramp AC is a segment of a cycloid (the "path of quickest descent") and the ramp BC is a straight line extended up slightly to D with a "return ramp" DA.

A ball released at A rolls down the ramp AC to C covering a greater distance in a shorter time that it would have had it rolled down BC to C. The relation *Velocity = Distance/Time* thus implies that the ball arrives at C with greater velocity than it would have had it rolled down BC. This added velocity enables the ball to roll from C up to *and past* B to a point D a little farther along. It then returns to A along the inclined ramp DA to repeat the cycle endlessly. ◇

Contributed by Eric Chandler of Randolph-Macon Woman's College in Lynchburg, VA.

5. The bouncing ball

Daniel J. Scully of St. Cloud University in Minnesota has encountered technology promoters explaining the use of a sonar sensing device that can be attached to a graphing calculator. The device tracked the height of a bouncing ball over time and the data collected were fed into the calculator to produce a graph similar to Figure 11.5a.

Using least-squares methods, formulae of the form $s(t) = -\frac{1}{2}gt^2 + v_0 t + s_0$ for the humps and an exponential formula fitting the peaks of the humps were obtained. This suggested, without any justification, that exponential de-

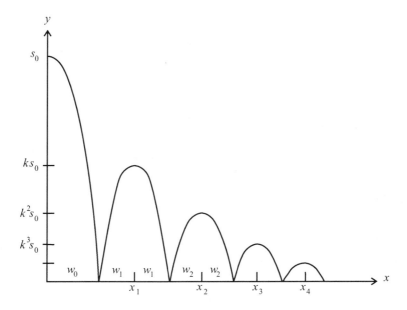

FIGURE II.5a

cay is the appropriate model for describing the declining peaks of the graph. Scully says:

> I object to the choice of the exponential-decay model. In a perfectly elastic collision, the kinetic energy of the system would remain constant through the collision, but in a less-then-perfectly elastic collision, a first-round approximation would assume that the kinetic energy would decrease by some fixed proportion. Simple calculations show that this translates to proportional decreases in the heights of successive bounces, after one neglects air resistance and assumes acceleration due to gravity is constant (both standard assumptions).
>
> The difficulty arises when we realize that in our graphs the units on the horizontal axis do not indicate the number of bounces: they indicate time (typically in seconds). Since the total elapsed time for all of the bounces is finite — a standard geometric series problem — any continuous curve that passes through all the peaks and extends over a closed interval must intersect the x−axis. This makes the exponential model inappropriate. What elementary curve passes through all the peaks of the graph? The following derivation shows that a parabola does.

Let s_0 be the initial height from which the ball is dropped and left to bounce. Assume that the height of each bounce is decreased from the previous one by a constant proportion k, $0 < k < 1$. We assume that the ball is released at time $x = 0$. Let y_n be the height of the nth bounce. Clearly, $y_n = s_0 k^n$. Let w_n equal the time between the ball's nth bounce and its following apex (as in the diagram below). Assuming that each hump is a parabola — hence symmetric around its vertex — if we let x_n equal the time it takes from the moment the ball is initially dropped until it reaches its apex after its nth bounce, we get $x_n = w_0 + 2w_1 + 2w_2 + \cdots + 2w_{n-1} + w_n$.

For each j, w_j equals the positive root of $-\frac{1}{2}gx^2 + y_j = 0$, so that $w_j = \sqrt{2y_j/g} = \sqrt{2s_0 k^j/g}$.

After working out the value of x_n, we find that the coordinates for the apexes $(x, y) = (x_n, y_n)$ of the parabolas are given by

$$x = \sqrt{\frac{2s_0}{g}}\left[\frac{2(1 - \sqrt{k}^{n+1})}{1 - \sqrt{k}} - (1 + \sqrt{k}^n)\right]$$

$$y = s_0 k^n.$$

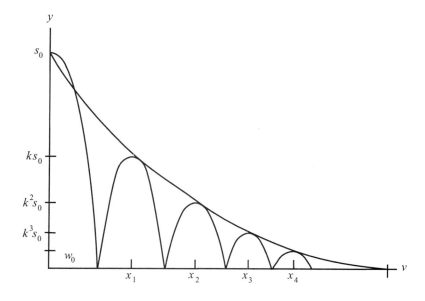

FIGURE II.5b

Using $k^n = y/s_0$ in the equation for x, solving for $\sqrt{y}$ and squaring yields

$$y = \frac{g}{2}\left(\frac{1 - \sqrt{k}}{1 + \sqrt{k}}\right)^2 x^2 - \sqrt{2gs_0}\left(\frac{1 - \sqrt{k}}{1 + \sqrt{k}}\right)x + s_0$$

whose graph is a parabola with a y−intercept of s_0 and vertex on the x−axis with abscissa $v \equiv \sqrt{2s_0/g}(1 + \sqrt{k})/(1 - \sqrt{k})$. This value v is the limit of x_n and represents the finite time required for the ball to do all of its bounces.

6. Watch your ears!

Problem. The loudness of the sound of a stereo speaker is inversely proportional to the square of the distance of the listener from the speaker.

a. Express the loudness L in decibels as a function of the distance d from the speaker.

b. When you are sitting 10 feet from the speaker, the loudness is 40 decibels. What is the loudness when you are 6 feet from the speaker?

Answer.   **a.** $L = L(d) = k/d^2$; $40 = k/10^2$, $L(10) = 40$, $k = 4000$. Therefore $L = 4000/d^2$. **b.** $L(6) = 4000/36 \sim 111.1$ decibels. ♡

This is a worked example from a recent intermediate algebra text. For reference, the sound of a quiet radio is about 40 decibels and of a riveter is 95 decibels; the threshold of pain is 120 decibels.

Contributed by Bruce Yoshiwara of Los Angeles Pierce College in Woodland Hills, CA.

7. Positive series with a negative sum

Consider this infinite matrix:

$$\begin{pmatrix} 1 & 1 & 1 & 1 & \cdots \\ 1 & \frac{1}{2} & \frac{1}{4} & \frac{1}{8} & \cdots \\ 1 & \frac{1}{3} & \frac{1}{9} & \frac{1}{27} & \cdots \\ 1 & \frac{1}{4} & \frac{1}{16} & \frac{1}{64} & \cdots \\ & & \cdots & & \end{pmatrix}$$

Summing up the row totals yields

$$(1 + 1 + \cdots) + (1 + 1) + (1 + \frac{1}{2}) + (1 + \frac{1}{3}) + \cdots$$

$$= 1 + 1 + 1 + \cdots + 1 + \frac{1}{2} + \frac{1}{3} + \cdots$$

$$= \sum_{k=1}^{\infty} 1 + \sum_{k=1}^{\infty} \frac{1}{k}$$

while summing over the column totals yields

$$\sum_{k=1}^{\infty} 1 + \sum_{k=1}^{\infty} \frac{1}{k} + \sum_{k=1}^{\infty} \frac{1}{k^2} + \sum_{k=1}^{\infty} \frac{1}{k^3} + \cdots.$$

Since both of these sum the terms in the matrix, they should be equal. Cancelling common terms yields

$$0 = \sum \frac{1}{k^2} + \sum \frac{1}{k^3} + \sum \frac{1}{k^4} + \sum \frac{1}{k^5} + \cdots$$

$$= \frac{\pi^2}{6} + \sum \frac{1}{k^3} + \frac{\pi^4}{90} + \sum \frac{1}{k^5} + \cdots$$

so that $\sum(1/k^3) + \sum(1/k^5) + \cdots$ must be negative.

Contributed by William A. Simpson of Michigan State University in East Lansing.

A centennial tribute to Sam Loyd

Dean Clark, Department of Mathematics, University of Rhode Island, Kingston

About one hundred years ago Sam Loyd, the "Prince of the Puzzle Makers," created his amazing "Get Off the Earth" puzzle. Martin Gardner [1], [2] called it Loyd's greatest creation and, unquestionably, it was sensational. It reportedly sold 10 million copies in its day but now it is a rare collector's item. I've brought Loyd's puzzle up to date in Figures 1(a) and (b), below. You cut out the small moon in (a) and rotate it 30° counterclockwise to obtain (b). One of the astronauts simply vanishes! Unlike Loyd's fiendish version, in which similar-looking Chinese warriors hovered over the earth, it is very easy to find which astronaut disappears.

There seems to be a good deal of misconception about this paradox and the linear version of it shown in Figure 2, below. The reader may have seen something like Figure 2 involving vanishing Leprechauns. According to Gardner [2], TV personality David Frost was so mystified by the Leprechauns that he featured the puzzle on his show and solicited audience analysis! Evidently,

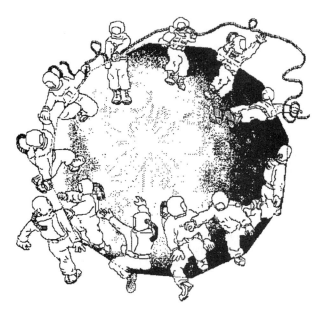

FIGURE l(a)

There are 12 astronauts in this picture.

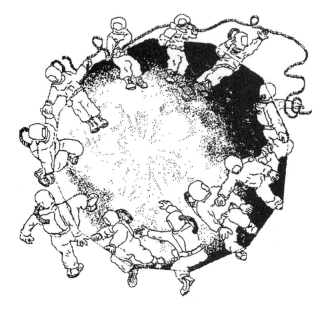

FIGURE l(b)

After the moon is cut out and rotated 30° counterclockwise there are 11 astronauts. Which one vanished?

FIGURE 2(a)

There are 16 faces (including a mask, which counts as a face) in this picture. Cut out pieces A and B and switch their position.

FIGURE 2(b)

Now there are 15 faces. Who's missing?

close to a century of human progress has not sharpened man-on-the-street responses to this strange phenomenon.

Is it mathematics? Consider that *discontinuous* point mappings (the cutting and switching in two dimensional Euclidean space) are essential in creating Figures 1 and 2, since topological connectivity is preserved under continuous transormation. The astronauts and faces are, after all, connected sets and it's hard to get them to vanish without introducing discontinuity.

Consider it as a counting problem: the combinatorics of the *Twilight Zone*. But even there, the *pigeonhole principle* remains valid. The boundaries of the astronaut-pictures define Venn-like sets which carry in their interiors *information* that identifies each astronaut. In Figure 1(b) there are 11 such set "carriers" and there are also 11 in Figure 1(a). But with 12 individuals in Figure 1(a), one carrier must contain *two* astronauts. The two can be seen trying to knock each other out of orbit at approximately 8 o'clock in Figure 1(a). This is a feature that had to be planned for in advance (Sam Loyd's original "Get off the Earth" also had it) in order to make the effect work.

What about *arithmetic* invariants? With respect to Figure 2, Gardner [1] asserts that the number of units in the two pieces to be switched must be relatively prime, but Figure 2 shows that this claim is false, leaving open the question of exactly *who* is missing in Figure 2(b)! Regrettably, space limitations prevent us from developing an answer.

In any case, I think Sam Loyd would have liked these pictures and would have liked knowing that his vanishing illusion continues to evade successful explanation after all these years.

References

1. Martin Gardner, *Mathematics, Magic and Mystery* (New York: Dover Publications, 1956)
2. ibid, *Wheels, Life and Other Mathematical Amusements* (New York: W.H. Freeman and Company, 1983)

References

Here is a sample of references that deal with mathematical paradoxes and fallacies. Journals often run items, either as short notes or space fillers, and you will find a few of these in this book. Many of the best fallacies and paradoxes found their way across the desk of Martin Gardner. These originally appeared in his *Mathematical Games* columns in *Scientific American*, which have been collected into many soft-covered volumes that are easy to obtain. I would be grateful to readers who provide other references.

V.M. Bradis, V.L. Minkovskii & A.K. Karcheva, *Lapses in mathematical reasoning*. Pergamon, Oxford, 1963.

Bryan H. Bunch, *Mathematical fallacies and paradoxes*. Van Nostrand, Reinhold, New York, 1982.

Stephen K. Campbell, *Flaws and fallacies in statistical thinking*. Prentice-Hall, Englewood Cliffs, NJ, 1974. (ISBN 0-13-322214-4)

Barry Cipra, *Misteaks and how to find them before the teacher does: a calculus supplement*. Second edition. Academic, HBJ, Boston, San Diego, 1989.

Richard J. Crouse & Clifford W. Sloyer, *Mathematical questions from the classroom*. Janson Publications, 1987. (ISBN 0-939765-04-7)

Ya. S. Dubov, *Mistakes in geometric proofs*. Heath, 1963.

Underwood Dudley, *Mathematical cranks*. Mathematical Association of America, 1992. (ISBN 0-88385-507-0)

Underwood Dudley, *The trisectors*. Mathematical Association of America, 1994. (ISBN 0-88385-514-3)

Glenn W. Erickson & John A. Fossa, *Dictionary of paradox.* University Press of America, Boston, 1980. (ISBN 0-7618-1065-X/0-7618-1066-8)

Nicholas Falletta, *The paradoxicon: a collection of contradictory challenges, problematical puzzles, and impossible situations.* John Wiley, New York, 1983, 1990. (ISBN 0-471-52950-8)

Daniel Fendel & Diane Resek, *Foundations of higher mathematics: exploration and proof.* Addison-Wesley, 1990.

Patrick Hughes & George Brecht, *Vicious circles and infinity: a panoply of paradoxes.* Doubleday, New York, 1975.

W. Lietzmann, *Wo stekt der Fehler?* Teubner, Stuttgart, 1953.

E.A. Maxwell, *Fallacies in mathematics.* Cambridge University Press, 1959, 1961 (paperback 1963).

Eugene P. Northrop, *Riddles in mathematics: a book of paradoxes.* Van Nostrand, New York, 1944.

T.H. O'Beirne, *Puzzles and paradoxes* Oxford UP, London, 1965.

William L. Schaaf, *A bibliography of recreational mathematics.* National Council of Teachers of Mathematics, 1970. (ISBN 0873530209) (There are several updates in the *Journal of Recreational Mathematics* over the period 1983–1987.)

I.F. Sharygin, So what's wrong? *Quantum* 8:6 (July/August, 1998) 34–37, 53–54.

Raymond H. Smullyan, *What is the name of this book? The riddle of dracula and other puzzles* Prentice-Hall, Englewood Cliffs, NJ, 1978.

Gábor J. Székeley, *Paradoxes in probability theory and mathematical statistics.* Akadémiai Kiadó, Budapest & D. Reidel, Dordrecht, Holland, 1986 (distributed by Kluwer).

Proceedings of the Second International Seminar, Misconceptions and educational strategies in science and mathematics. Cornell University, 1987.

Index of Topics

Index of Names